EINSTEIN IN OXFORD

EINSTEIN IN OXFORD

Andrew Robinson

BODLEIAN
LIBRARY
PUBLISHING

To James

First published in 2024 by Bodleian Library Publishing
Broad Street, Oxford OX1 3BG
www.bodleianshop.co.uk

ISBN: 978 1 85124 638 0

Publisher: Samuel Fanous
Managing Editor: Susie Foster
Editor: Janet Phillips
Picture Editor: Leanda Shrimpton
Cover design by Dot Little at the Bodleian Library
Designed and typeset by Laura Parker in Kingfisher in 9 on 12.94 font
Printed and bound in China by C&C Offset Printing Co., Ltd on 120 gsm Chinese Baijin
Pure paper

British Library Catalogue in Publishing Data
A CIP record of this publication is available from the British Library

CONTENTS

Foreword

The inspiration for Albert Einstein's special theory of relativity – formulated in Switzerland and published in Germany in 1905 – was unquestionably British physics. On the walls of his apartment in 1920s Berlin, and later in his American house in Princeton, Einstein hung portraits of three British natural philosophers: Isaac Newton, Michael Faraday and James Clerk Maxwell – and no other scientists.

'England has always produced the best physicists', he said in 1925 to his pupil Esther Salaman in Berlin. He advised that she should study physics at the University of Cambridge: the home of Newton and of Einstein's ally Arthur Eddington, the astrophysicist whose team had proved the 1915 general theory of relativity through observing the 1919 solar eclipse. As he explained to her: 'I'm not thinking only of Newton. There would be no modern physics without Maxwell's electromagnetic equations: I owe more to Maxwell than to anyone.' Without exaggeration, we can say that Britain is the country that made Einstein the worldwide phenomenon he is known as today.

When he first visited Britain in 1921 he gave celebrated lectures in Manchester and London and also visited Oxford, with his wife Elsa, for a few hours at the invitation of Frederick Lindemann, the university's professor of experimental philosophy (that is, physics) and director of its Clarendon Laboratory. Soon Lindemann began wooing his fellow German physicist to return for a longer period. After much global travelling in the 1920s, Einstein stayed for some weeks in Oxford as Lindemann's guest in 1931, 1932 and 1933 – latterly as a refugee from Nazi Germany. He gave public scientific lectures – including one of his most controversial ever, 'On the Method of Theoretical Physics' – and received an honorary doctorate. But he also took part in the cultural, political and social life of both the university and the city, including playing a borrowed violin with musicians at the Oxford Chamber Music Society, speaking to Quakers about the importance of pacifism, and sailing on the River Thames in a skiff with an Oxford research student. While staying at Lindemann's college Christ Church, in rooms that were once occupied by Lewis Carroll, author of *Alice's Adventures in Wonderland*, Einstein inadvertently provided Oxonians with 'Adventures in Einstein-land' – as noted by Andrew Robinson in this volume: the first book to be devoted solely to Einstein in Oxford.

Today, the best known of these adventures pivots around Einstein's blackboard from his 1931 Rhodes House lectures, on public display in Oxford's History of Science Museum. Its mathematical chalk markings have always attracted curious, and often reverential, visitors from all over the world. With its markings very nearly erased in the usual way with blackboards, it was rescued by some admiring Oxford science dons just after the lecture as a unique relic of a genius. This annoyed Einstein, who was always modest about his legend and unwilling to provoke envy among fellow scholars. 'Not even a cart-horse could endure so much!' he wrote in his private diary. Perhaps he also realized that the blackboard's mathematics contained a calculation error by him about the expansion of the Universe!

In a Carroll-like thank-you poem preserved in the Bodleian Library – written by Einstein in rhymed German to an absent Christ Church classical scholar whose rooms he had borrowed in 1931 – he compared himself, both humorously and seriously, to a 'barbarian' roaming the planet. The unique mingling of science and humanity, formality and informality in his relationship with Oxford reveals another fascinating, less familiar, episode of the Einstein story.

Silke Ackermann
Director, History of Science Museum

Preface
EINSTEIN'S OXFORD BLACKBOARD

I rejoice at the new universe to which [Einstein] has introduced us. I rejoice in the fact that he has destroyed all the old sermons, all the old absolutes, all the old cut and dried conceptions, even of time and space, which were so discouraging...

[speech by George Bernard Shaw, 1930]

Albert Einstein visited Britain in October 1930 from his home in Germany to attend a charitable fund-raising dinner for disadvantaged East European Jews, given in his honour in London. There he heard George Bernard Shaw's words of praise quoted above – part of a speech 'renowned as one of the finest 20th-century examples of a public eulogy', according to a recording of it reissued by the British Library in 2005 to celebrate the centenary of Einstein's special theory of relativity. The great dramatist and writer clearly placed the great physicist in the pantheon of the immortals, while admitting that he himself had been unable to understand relativity fully, despite his best efforts. According to Shaw, Einstein belonged among the 'makers of universes', in the company of Pythagoras, Ptolemy, Aristotle, Copernicus, Kepler, Galileo and Newton – 'not makers of empires' like Alexander and Napoleon. 'And when they have made those universes, their hands are unstained by the blood of any human being on earth.'

Einstein's response in German mixed genuine humility with typical humour. 'I, personally, thank you for the unforgettable words which you have addressed to my mythical namesake, who has made my life so burdensome', yet who, 'in spite of his awkwardness and respectable dimension, is, after all, a very harmless fellow'.

About six months later, in May 1931, Einstein's legend accompanied him to Oxford. He went there as a guest of the college Christ Church, to give three lectures on relativity, cosmology and his latest work in physics at Rhodes House, and to receive an honorary Oxford doctorate at the Sheldonian Theatre. While lecturing, Einstein naturally used equations and diagrams chalked on several blackboards. The second lecture required 'two blackboards, plentifully sprinkled beforehand in the international language of mathematical symbol', as *The Times* reported. They described the density, size and age of the expanding universe, which Einstein calculated to be ten billion years old.

Today, one of those blackboards is the most famous object of the 18,000 or so objects in Oxford's History of Science Museum, located close to the Sheldonian Theatre and the Bodleian Library. Visitors from all over the world come to the museum especially to view the Einstein blackboard. It is a 'relic of a secular saint', as the museum's website ironically describes it. 'Some visitors today treat it almost as an object of veneration, anxiously requesting its location on arrival and eager to experience some connection with this near-mythical figure of science.'

The idea of rescuing and preserving Einstein's blackboards seems to have come from some Oxford dons who attended his first lecture on 9 May. A memo from the then-warden of Rhodes House to one of the Rhodes trustees, dated 13 May, states plainly that:

Some of the scientists seem to be anxious to secure for preservation in the Museum the blackboard upon which Einstein draws. I was first approached about it by de Beer, who is a fellow of Merton [College],

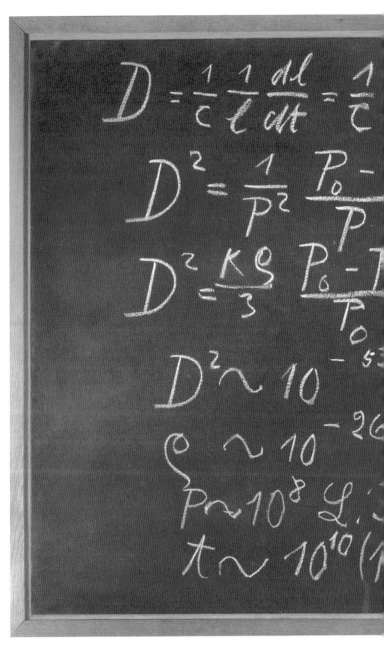

$$D = \frac{1}{c}\frac{1}{\ell}\frac{d\ell}{dt} = \frac{1}{c}$$

$$D^2 = \frac{1}{P^2}\frac{P_0 -}{P}$$

$$D^2 = \frac{\kappa\varrho}{3}\frac{P_0 - }{P_0}$$

$$D^2 \sim 10^{-53}$$

$$\varrho \sim 10^{-26}$$

$$P \sim 10^8 \quad \mathcal{L} :$$

$$t \sim 10^{10} (1$$

1. Blackboard used by Einstein at Rhodes House, Oxford, May 1931. Preserved against his wishes by Oxford dons, it is now kept in the History of Science Museum, Oxford.

$$\frac{dP}{dt}$$

$$\sim \frac{1}{P^2} \qquad (1a)$$

$$\cdots + \kappa \rho \qquad (2a)$$

$$\gamma$$

and now Gunther has written to me, asking whether, if it is desired, the blackboard with Einstein's figures on it may be given to the university.

Gavin de Beer was an embryologist, who became a fellow of the Royal Society and director of the Natural History Museum in London. Robert Gunther was a zoologist and historian of science, who in 1924 created the nucleus of what became the Museum of the History of Science in Oxford in 1935. Yet another Oxford academic involved in the rescue was the chemist Edmund Bowen, also a fellow of the Royal Society, whose laboratory work in photochemistry had confirmed some of Einstein's theoretical work. Presumably, they were among the audience on 16 May for the second Rhodes House lecture. Certainly, on 19 May, Gunther formally thanked the secretary of the Rhodes trustees for 'your present' to the newly established museum 'of two blackboards used by Professor Einstein in his lecture'. Although one of these blackboards was later accidentally cleaned in the museum's storeroom, the other one survived, to be venerated in our time.

Einstein, had he come to know of this accidental erasure of his work, would certainly have laughed, as he loved to do right until the end of his life in 1955. In May 1931, he was firmly opposed to preserving his blackboards – firstly because he regarded them as mere ephemera, secondly because they showed work in progress that was almost certain to be superseded by his subsequent calculations, and lastly because their preservation drew invidious attention to his legendary status. Indeed, on 16 May, after the second lecture, Einstein told his diary with singular annoyance:

The lecture was indeed well-attended and nice. [But] the blackboards were picked up. (Personality cult, with adverse effect on others. One could easily see the jealousy of distinguished English scholars. So I protested; but this was perceived as false modesty.) On arrival

THE LEWIS EVANS COLLECTION OF
SCIENTIFIC INSTRUMENTS

The Old Ashmolean
Broad Street
Oxford

19 May 1931

Dear Sir,

Your Present of 2 black boards
used by Professor Einstein in his lecture
has been duly received, and I beg you to accept
my sincere Thanks for your kindness in thus
contributing to promote the objects for which the
Lewis Evans Collection of Scientific Instruments
has been founded in the University.

I am,

Yours faithfully,

R. T. Gunther
Curator.

To the Secretary
The Rhodes Trustees.

2. Letter from historian of science Robert Gunther to the Rhodes trustees, Oxford, May
1931, thanking them for donating two blackboards used by Einstein in his second Rhodes
House lecture.

[back in Christ Church] I felt shattered. Not even a cart-horse could endure so much!

In a way, the whole episode of the blackboards – mingling as it did genuinely warm appreciation of the man with relatively superficial understanding of his thought – is typical of Einstein's experience of Oxford, on his three residences in the city in 1931, 1932 and 1933. While he clearly enjoyed his personal contact with scientists and a wide range of non-scientists in Oxford, including artists and musicians, he did not much care either for his legendary status or for the formalities of college and university life, which generally amused and occasionally irritated him. In his diary, he mocked the dinner-jacketed and gowned dons dining in the ceremonial splendour of Christ Church Hall as 'the holy brotherhood in tails' ('*der heiligen Brüderschaar im Frack*'). He always disliked having to wear formal dress – and even, famously, socks. In a thoughtful and witty poem (now in the Bodleian Library) that he wrote to thank the absent Christ Church don whose college rooms he borrowed in 1931, Einstein called himself, only partly in jest, an old 'hermit' and a 'barbarian' on the roam.

As the story of Einstein in Oxford told in this companion will reveal, he always stayed true to himself – wherever he roamed, whether it was Britain, Germany, Japan, Palestine, South America or the United States, in which he spent the final two decades of his life. And this, most probably, is why Einstein still appeals so strongly beyond the world of scientists to the entire globe.

Chapter 1

OXFORD'S ADVENTURES IN EINSTEIN-LAND

Einstein's first visit to Oxford took place in June 1921, during his first visit to Britain, hosted in London by Lord Richard Haldane, a leading Liberal politician keen to build bridges with Germany after the horrors of the First World War. Einstein's host in Oxford was the physicist Frederick Alexander Lindemann, professor of experimental philosophy (that is, physics) and director of the Clarendon Laboratory at Oxford, who had first met Einstein in 1911 in Brussels at the famous Solvay Congress of the world's greatest scientists. Here the up-and-coming Einstein had given a key speech and the still-unknown Lindemann had been the secretary of the Congress. On 14 June 1921, Lindemann whisked Einstein and his wife Elsa away from London by car, in order to give them a first glimpse of Oxford. An announcement in *The Times* on 15 June stated: 'Professor Einstein paid a private visit to Oxford yesterday as the guest of Dr Lindeman [*sic*], of Wadham College. A tour was made of the principal University buildings, and the Professor returned to London in the evening.'

Unfortunately, neither Einstein nor Lindemann recorded any impressions of the visit. It must have been brief, since the Einsteins in fact returned to London earlier than mentioned, by an afternoon train. Presumably Lindemann, probably at Einstein's request, did not introduce him to any of his colleagues and friends so that the Einsteins could have enough time to see the buildings. They did

not even meet the university's vice-chancellor. As Lindemann commented in a scribbled letter from Oxford to Haldane: 'I enclose a note [for Einstein] which the vice-chancellor just brought here. He said he much regretted not having known of Einstein's visit in time to offer him some hospitality.' To which Haldane responded the following day, after talking to Einstein in London: 'They both enjoyed greatly their visit to Oxford and to yourself.' Einstein seems to have agreed. A few days after returning to Germany on 17 June, he wrote to thank Haldane with transparently sincere emotion: 'The wonderful experiences in England are still fresh in my mind and yet like a dream. The impression this land with its wonderful intellectual and political tradition left on me was a profound and lasting one, even larger than I had expected.'

Lindemann would be the key figure in Einstein's relationship with Oxford in the 1930s, so it is worth giving some background about his life and personality. Though born in Germany (in 1886, seven years after Einstein), the son of a decidedly wealthy German father and an American mother, Lindemann had been brought up and educated as a British citizen. Indeed, all his life he resented the accident of his birthplace and came to regard himself as more English than the English, with an accompanying distrust for certain aspects of Germany. In his mid-teens, however, he had returned to Germany for further schooling and then attended the University of Berlin. After graduating, he earned a doctorate in physics. Just before the outbreak of war in August 1914 Lindemann left Germany to avoid being interned, and settled for good in England. In 1915, after failing to obtain a military commission because of his German background (a rejection which rattled him), he joined the Royal Aircraft Factory at Farnborough, learned to fly the following year,

3. Einstein with Richard Haldane, London, 1921. Haldane, as a leading Liberal politician, invited Einstein to Britain as part of an effort to build bridges with Germany after the First World War.

4. Frederick Lindemann, Einstein's host in Oxford in 1931, 1932 and 1933. A physicist of note, also much involved with British politics, he had a complex relationship with his guest.

and soon became semi-legendary for having extricated himself from a potentially lethal aircraft spin through rapid mental calculation, navigational skill and sheer courage, while empirically testing his own theory to explain the nature of the spin. Appointed a professor in Oxford in 1919, he soon joined Christ Church, in 1922. There he would reside in a suite of fine rooms until his death in 1957, soon becoming known as 'The Prof.', and, after 1941, as Lord Cherwell, the scientific adviser and confidant of Prime Minister Winston Churchill.

As a physicist, Lindemann was highly rated, but never placed in the top rank. 'He was a man of intuition and flair in widely diverse fields, but he never pursued any one subject long enough to become its complete master. Much of his brilliance was shown in discussion at scientific conferences, and has not survived in published form', commented historian Robert Blake, a Christ Church colleague, in the *Oxford Dictionary of National Biography*. 'For this reason later generations have not found it easy to understand the high esteem in which he was held by such persons as Albert Einstein, Max Planck, Max Born, Ernest Rutherford and Henri Poincaré.' This assessment is confirmed by Einstein's own private summary of Lindemann, as reported by another Christ Church colleague, economist Roy Harrod:

The Prof., so it went, was essentially an amateur; he had ideas, which he never worked out properly; but he had a thorough comprehension of physics. If something new came up, he could rapidly assess its significance for physics as a whole, and there were very few people in the world who could do that.

As a politician, Lindemann's views were well to the right (though never sympathetic to Nazism). 'He was an out-and-out inequalitarian who believed in hierarchy, order, a ruling class, inherited wealth, hereditary titles and white supremacy (the passing of which he regarded as the most significant change in the twentieth century)', wrote Blake. Yet in private he was kind-hearted and most generous to those in need, drawing on his wide contacts and personal wealth. Such attitudes – both public and private – underscored Lindemann's obituary of Einstein for the *Daily Telegraph* in 1955. Overflowing with respect for Einstein's science, it was not surprisingly somewhat critical of his liberal and pacifist politics: 'Like many scientists Einstein was politically rather naïve. He hated violence and war and could not understand why his own natural sweet reasonableness was not universal. Absolutely truthful himself, he tended to be credulous in political questions and was easily and often imposed on by unscrupulous individuals and groups.' Yet, Lindemann concluded: 'As a man his simplicity and kindliness, his unpretentious interest in others and his sense of humour charmed all who knew him.'

Undoubtedly, Lindemann's own contrary personality polarized his contemporaries (as it does even today). 'It has often been asked how a prickly, eccentric, arrogant, sarcastic and uncooperative man – to use some of the adjectives from time to time levelled against Lindemann – could have developed and sustained such a warm friendship with Churchill', according to Adrian Fort, Lindemann's most recent biographer. 'The answer is of course that he did not display those characteristics to Churchill.' Presumably the same was true in Lindemann's somewhat less warm, and certainly less intense, friendship with Einstein.

It was in 1927 that Lindemann began to court Einstein for a second visit to Oxford. He had the support of the Rhodes Trust, who wished to launch the Rhodes Memorial Lectures in Oxford in memory of Cecil Rhodes, the British-born Victorian businessman, mining magnate and politician in southern Africa, whose strong support for imperialism and white supremacy would presumably have appealed to Lindemann – if rather less to Einstein (and not at all to most Oxford dons today). The trustees' aim was to attract to Oxford leading figures in public life, the arts, letters, business or science from around the world, whose presence would counteract the prevailing insularity of the university. To cite the devastating words of a British government commission of enquiry into the universities at Oxford and Cambridge, reporting in 1922: 'It is a disaster that, at a moment when we have become far more deeply involved than ever before in the affairs of countries overseas, our highest academical class is condemned through poverty to knowing little or nothing of life or learning outside this island.'

Although the Rhodes trustees were conscious of the recondite nature of relativity, and wary of the fact that Einstein would need to speak in German, they pressed ahead with an invitation, given his worldwide renown. One of them, the Oxford historian and Liberal politician H.A.L. Fisher, recruited Lord Haldane to make the introductory approach. Haldane wrote to Einstein in Berlin in June 1927, introducing the unfamiliar Rhodes lectures: 'The university and the trustees desire that the lectures should next year be delivered by the foremost man of science in the world, and they are unanimous in their choice of your name'. Haldane hoped Einstein would accept, not least because this would be 'very good for Anglo-German relations that the choice should be proclaimed to the world'. As for the subject, it should be 'just what you select. Not too technical in detail, but extending to anything you please, mathematico-physical or otherwise'. As for the Oxford audience, it would include 'learned men as well as the public'. At the end of his letter Haldane mentioned that Lindemann would soon get in touch with the details of the invitation.

Einstein was interested, but he refused, for a mixture of reasons. He frankly explained to Lindemann in July: 'How gladly would I accept, particularly as I value highly the milieu of English intellectuals, as being the finest circle of men which I have ever come to know.' Unfortunately, however, scientific commitments to people in Germany would prevent him from being away for such a long time. Secondly, his poor health would make 'a long stay in foreign and unfamiliar surroundings … too great a burden for me, particularly bearing in mind the language difficulty'. Lastly, he modestly confessed that his current work was not at the forefront of physics, as compared with that of some other physicists who would appeal more to an Oxford audience.

But in August Einstein changed his mind, and gave Lindemann encouragement: 'During the holidays I have often reproached myself because I haven't accepted your kind invitation to Oxford.' Perhaps he could come to the university for just four weeks during the next summer term? 'It is very important to me that in England, where my work has received greater recognition than anywhere else in the world, I should not give the impression of ingratitude.' However, he realized that following his earlier refusal someone else had probably been invited in his place. In which case, he trusted that Lindemann would make clear to the Rhodes trustees the warmth and gratitude he felt for their proposal.

By now, the American educationist Abraham Flexner (later to be a key figure in Einstein's life in the United States) had been approached to give the 1928 Rhodes lectures. Einstein was therefore invited to speak in the following year, 1929, and apparently accepted; but again negotiations broke down for reasons of health. In the meantime, the Rhodes lectures began to establish themselves after a well-attended series on world politics given in 1929, not at Rhodes House but at the nearby larger Sheldonian Theatre, by Jan Christiaan Smuts, the South African soldier and statesman (with racial views similar to those of Rhodes). This success renewed the determination of both the trustees and Lindemann to secure agreement from Einstein. At last, following a personal visit to Einstein in Berlin by Lindemann in October 1930, arrangements were finalized. For a lecture fee of £500, Einstein agreed to visit Oxford for some weeks in May 1931, to give the Rhodes lectures, and to live in Lindemann's college, Christ Church – this time without his wife Elsa, who remained in Berlin.

The opening lecture by Einstein took place in the Milner Hall at Rhodes House on 9 May 1931. Given in German (like his other two lectures) without notes but with a blackboard, its English title was simply 'The Theory of Relativity'. The second lecture, in the same place on 16 May, dealt with relativity and the expanding universe: a subject then in a state of great scientific flux following the discoveries of the American astronomer Edwin Hubble in the late 1920s. It required the two blackboards already mentioned in the Preface. The last lecture, also in Rhodes House on 23 May, immediately after the university had awarded Einstein an honorary doctorate in the nearby Sheldonian Theatre, tackled Einstein's constantly evolving unified field theory: 'an account of his attempt to derive both the gravitational and electromagnetic fields by the introduction of a directional spatial structure', as the scientific journal *Nature* chose to announce it.

The content of the lectures was of no lasting scientific significance, since it either repeated Einstein's existing published work on relativity or was quickly rendered redundant by his (and others') subsequent ideas. More interesting is the reaction of the very mixed Oxford audience, which included some 500 selected students, to such an unparalleled educational-cum-social occasion.

The *Oxford Times* captured the atmosphere in two reports on the opening and final Einstein lectures. The first of these, headlined 'Women and Relativity', remarked on 15 May about the challenge to the audience of combining German with physics:

Women in large numbers flocked to hear Prof. Einstein speak on relativity at Rhodes House on Saturday morning. The front of the hall was filled with heads of houses and the back of the hall and the gallery with younger members of the university. It was unfortunate that no interpreter was provided, but Oxford seems to fight shy of interpreters. One wonders how many of those who were present thoroughly understood German, or if they could understand the language in which Prof. Einstein spoke, how many of them could follow the complexities of relativity. Prof. Einstein is a man of medium height with a wealth of black curly hair, already greying. Entirely unaffected, he had charm as well as simplicity of manner, which appealed to his audience.

Although most of the audience stayed the course, at least one academic, a professor of mathematics, slipped out after five minutes because he was unable to follow German.

The second report on 29 May began with a reference to the just-completed doctoral ceremony in the nearby Sheldonian Theatre:

Prof. Einstein, wearing his new doctor's robes, acknowledged the applause which greeted his appearance by smiling and bowing. His manner in beginning his lecture suggested that he was dealing with a difficult part of the subject, and at first he spoke earnestly from the desk, with his hands clasped in front of him, only leaving it occasionally for the blackboard. As the lecture proceeded, not only equations but a singular diagram appeared on the blackboard, and Prof. Einstein gesticulated helpfully in curves with the chalk to explain it. At this point he turned repeatedly from his audience to the board and back. Later, the diagrams were rubbed off in favour of more formulae, and the better informed members of the audience were kept busy taking them down.

By now, at least one less-informed member sitting in the first row of the audience had fallen blissfully asleep, as Einstein later noted in his diary. The dean of Christ Church, Henry Julian White, a biblical scholar in his seventies, was understandably baffled by mathematical physics. Einstein was amused, and perhaps also learnt a lesson. For after one of the lectures, he apparently remarked in his curious English that the next time he had to lecture in Oxford, 'the discourse should be in English delivered'. Hearing this remark, one of his donnish companions, the physiologist John Scott Haldane (brother of Lord Haldane), was heard to murmur in German the ironic warning: '*Bewahre!*'

5. Einstein during his doctoral ceremony near the Sheldonian Theatre, Oxford, May 1931. He is following the university bedel (mace-bearer).

However, Einstein did follow his advice to himself. When he gave his most important lecture in Oxford, the Herbert Spencer Lecture on 10 June 1933, he had it translated into fluent English. Before beginning, he thanked three colleagues at Christ Church – a philosopher, Gilbert Ryle; a classicist, Denys Page; and a physicist, Claude Hurst – 'who have helped me – and perhaps a few of you – by translating into English the lecture which I wrote in German'. (Oddly, he did not mention a role for the bilingual Lindemann.) This time at Rhodes House, unlike in 1931, the audience did not have to struggle with an unfamiliar language – although Einstein's inimitably accented English, as he read from the dense translation, might sometimes have appeared to have been a foreign tongue. Nor were there any blackboards chalked with baffling mathematics – unlike in his three 1931 lectures – to be saved by eager science dons. That said, the Herbert Spencer Lecture was demanding enough to challenge any listening physicist or philosopher (as will be discussed in Chapter 2), not to mention the anonymous correspondent of *The Times* with the unenviable task of trying to summarize its content for a general readership. The *Times* report on 12 June was safely, if unimaginatively, headlined 'Basic Concepts in Physics'.

Like the 1931 lectures, Einstein's doctoral ceremony was not without comedy. It took place in the grandeur of the Sheldonian Theatre, designed by Sir Christopher Wren, in the presence of Oxford's vice-chancellor, Frederick Homes Dudden, a theological scholar and chaplain to King George V – and of course in front of a packed house, at least some of whose members were by now personally known to Einstein. Oxford's public orator, who presented the academically attired Einstein in Latin, had perhaps the most difficult role. He was a classical scholar, Arthur Poynton, later to be the master of University College, Oxford, with as little understanding of theoretical physics as Dean White of Christ Church.

The Latin oration opened with a reference to the solar eclipse in late May 1919 that had confirmed the general theory of relativity, almost exactly twelve years before, and the fact that 'Mercurius' (the planet Mercury) had on that occasion been observed in the position predicted by Professor Einstein. '*Atque utinam Mercurius hodie adesset, ut, cuius est eloquentiae, vatem suum laudaret!*' ('If Mercury were present today, he would of course praise his poet with his own eloquence!') At the end, Poynton attempted to relate Einstein's relativity to classical philosophy. According to a translation published a few days later in the *Oxford Times*:

> The doctrine which he interprets to us is, by its name and subject, interpreter of a relation between heaven and earth. It bids us view, under the aspect of our own velocity, all things that go on in space; to right and left, upward and downward, backward and forward. This doctrine does not in any way supersede the laws of physicists, but adds only the 'momentous' factor which they most desired. But it directly affects the highest philosophy, and it is not unwelcome to Oxford men, who have not the Euclidean temper of mind, but have learnt from Heraclitus that no man can step twice into the same river – nay, not even once; who are glad to believe that the Epicurean 'swerve' is not a puerile fiction; who, finally, in reading the *Timaeus* of Plato, have felt the want of a mathematical explanation of the universe more self-consistent and more in agreement with realities. This explanation has now been brought down to men by Prof. Albert Einstein, a brilliant ornament of our century.

Laudatory as it was, the oration contained no attempt to translate the term 'relativity' into Latin, no reference to gravity or electromagnetism, and not even a name check for the immortal – if Cambridge-based – Isaac Newton. (Contrast *The Times* in its lengthy editorial on 25 May, 'Professor Einstein at Oxford', which noted that: 'Like Newton, Professor Einstein is not primarily an

LADY MARGARET HALL

PHILIP MAURICE DENEKE LECTURE

Dr. Albert Einstein

will lecture on

EINIGES ZUR ATOMISTIK

at

Lady Margaret Hall

ON TUESDAY, 13 JUNE, at 8.30 P.M.

(Admission through new building entrance.   No tickets are required)

6. Poster for Einstein's Deneke Lecture at Lady Margaret Hall, the Oxford college of Margaret Deneke, June 1933. Its physics went over her head, as she freely admitted in her diary.

experimental physicist, but a mathematician.') Moreover, contrary to the oration, general relativity does indeed 'supersede the laws of physicists', in the sense that it is more fundamental than Newton's laws of motion. No wonder Einstein noted in his diary in the evening that the oration was 'serious, but not wholly accurate'. He must have based this remark on a translation given him during the day (perhaps by Lindemann), since Einstein did not understand Latin.

Yet, when the public orator was speaking, Einstein had recognized his mention of Mercurius. 'I had noticed his face lit up when "Mercury" was named', observed a friend in the audience at the Sheldonian. She was Margaret Deneke, who had got to know Einstein soon after his arrival in Oxford. Not only was Deneke fluent in German, she was also intensely musical: two characteristics that immediately endeared her to Einstein (as we shall see in Chapter 3). She persuaded a reluctant Einstein to give one final scientific lecture in Oxford: the Deneke Lecture, in memory of her musicologist father, at Lady Margaret Hall on 13 June 1933, with neurophysiologist Charles Sherrington, a Nobel laureate, in the chair. It was packed 'and many of our best friends failed to get seats', reported Deneke in her diary. Its subject was atomic theory, rather than relativity. 'While Dr Einstein was speaking and using his blackboard I thought I understood his arguments.' But, 'When someone at the end begged me to explain points, I could reproduce nothing. It had been the Professor's magnetism that held my attention'. This is perhaps an appropriate, if unintended, summary of Oxford's adventures in Einstein-land, other than for a tiny handful of its physicists.

Chapter 2

PHYSICS, MATHEMATICS AND THE NATURE OF REALITY

According to Lindemann, Einstein 'threw himself into all the activities of Oxford science, attended the colloquiums and meetings for discussions and proved so stimulating and thought provoking that I am sure his visit will leave a permanent mark on the progress of our subject'. This was written in June 1931 after Einstein's departure from Oxford. 'Combined with his attractive personality, his kindness and sympathy have endeared him to all of us and I have hopes that his period as Rhodes lecturer may initiate more permanent connections with this university which can only prove fertile and advantageous in every respect.'

Admittedly, Lindemann addressed this encomium to the secretary of the Rhodes trustees, Lord Lothian, probably with an eye on future funding for Einstein in Oxford. Nevertheless, Lindemann's comments were essentially true concerning Einstein's attitude. What they overlooked, however, was the fundamentally experimental orientation of Oxford science in 1931, which included hardly any theoreticians capable of discussing physics and mathematics at Einstein's level. There was to be no Oxford equivalent for Einstein of Cambridge's Arthur Eddington, the astronomer who had led the 1919 solar eclipse observations that validated Einstein's general relativity.

Thus, Einstein's dinner at New College with John Sealy Townsend, Wykeham Professor of Physics, led nowhere, other than a visit to the college chapel for a fine performance of Mozart's *Requiem* and a tour of Townsend's laboratory. And a lunch at Wadham College with the eminent pure mathematician, G.H. (Godfrey Harold) Hardy, was noteworthy only for Hardy's showing Einstein an interesting game with matches. The most promising candidate for a serious discussion was a cosmologist, Edward Arthur Milne, who had been appointed the first Rouse Ball Professor of Mathematics at Oxford in 1929. On 5 May 1931, Einstein attended a meeting with Milne at Trinity College, during which Milne gave a speech on novae (new stars) to a small group. He is a 'very clever man', Einstein noted in his diary. The following day Milne had further discussions with Einstein in the company of Lindemann, and another meeting with Einstein the day after that. But during 1932 Milne went on to develop a theory of 'kinematical relativity', based on his conviction that general relativity (published by Einstein in 1915–16) was unsound. It aimed to derive cosmological models by extending the kinematical principles of special relativity (Einstein's original theory of 1905). However, Einstein regarded Milne's approach as unsound. His doubts were confirmed by two independent analyses in 1935 and 1936, which 'showed that Milne's distinctive approach led, ironically, to the same set of basic models already studied in relativistic cosmology', according to *The Cambridge Companion to Einstein*.

Given the fluid state of cosmology and of his unified field theory in 1931, Einstein showed almost no interest in preparing his Rhodes lectures for publication – unlike Abraham Flexner and Jan Smuts, whose lectures were published by Oxford University Press in 1930. The secretary of the delegates of the press, R.W. Chapman, regarded Einstein's lectures – however demanding their subject matter might be – as a potential ornament to the publisher's list, and strongly pursued the possibility of publication. But Chapman received no reply from Einstein to his follow-up letters and cables. After a final proposal (apparently suggested by Lindemann) –

that Einstein might produce a reduced text of about 50 pages –
went nowhere, an exasperated Chapman gave up on what he now
called '*l'affaire* Einstein'.

Two years later, at a social gathering in Oxford in 1933, Einstein
intimated to the warden of Rhodes House, Carleton Allen, that
publication was impossible because 'he had since discovered
that everything he had put forward in the lectures was untrue'.
He explained, with 'rather comic contrition', that 'in my subject
ideas change very rapidly'. 'I had not the hardihood to say that
that was true of most subjects', commented Allen, when reporting
his conversation with Einstein in a letter to Lothian. Instead,
'I suggested that he might publish the lecture with a short note at
the end – "I do not believe any of the above" – or words to the effect.
He felt, however, that others might take up his ideas and convince
him that what he knew to be untrue was true'. As for Einstein's
vague suggestion that he might write an alternative book, Allen
told Lothian ironically that perhaps an Einstein book on 'My View
of Hitler' or 'Hitler in Time and Space' might help to recoup the
large amount that the Rhodes Trust had spent on Einstein. In the
end, his Rhodes lectures went entirely unrecorded, apart from an
eight-page pamphlet printed in early May 1931, probably compiled
with Lindemann's help, and three non-technical summary reports
published in *The Times*, apparently based on this pamphlet.
But Oxford University Press did at least get to publish Einstein's
much more significant 1933 Herbert Spencer Lecture.

Einstein's surviving Oxford blackboard provides interesting
insight into his difficulty with publication in 1931. In the history
of cosmology, the period 1917–32 is reminiscent of the early
seventeenth century, when Galileo Galilei's telescope provided
astronomical evidence for the Copernican solar system and
Johannes Kepler reformulated the planetary orbits as ellipses.
In this period, 'Theory fed on observation, observation fed on
theory, and in the end science ended up much grander and more

ON THE METHOD OF THEORETICAL PHYSICS

BY

ALBERT EINSTEIN

The
Herbert Spencer Lecture
DELIVERED AT OXFORD
10 JUNE 1933

OXFORD
AT THE CLARENDON PRESS
1933

7. Einstein's most significant scientific lecture in Oxford, 'On the Method of Theoretical Physics', published in 1933. Still controversial, it claimed precedence for mathematics over physics as a creative principle.

powerful than before. This time around, astronomers recognized the Milky Way as one of countless galaxies strewn through a vast and dynamic universe, each one composed of many billions of stars', wrote a former NASA space scientist, Corey Powell, in his study of Einstein, cosmology and religion. 'And the cosmologists were ready to make sense of it all, to demonstrate they could explain our place among the fleeting galaxies as readily as their ancestors had put the Earth in motion among the planets.'

At the beginning, in 1917, Einstein had set about applying general relativity to the universe as a whole. Comparatively uninformed about the latest empirical work in astronomy, his primary interest was not to construct a cosmological model from observations, but rather to build 'a spacious castle in the air' (as he told the astronomer Willem de Sitter), in order to test his theory. Much to his surprise, and even dismay, general relativity predicted the cosmos to be dynamic, not static: either expanding or contracting over time. Since Einstein knew of no astronomical evidence for such movement, he added a term to his existing field equations of relativity, intended to counteract the attractive influence of gravity and thereby stabilize the universe, making it static and therefore eternal, as he thought it should be. Known as the 'cosmological constant', it allowed Einstein to predict a cosmos that was both static and finite, with a radius and average density of matter that could be calculated from first principles rather than from astronomical observations. But he admitted in his published paper that the new term was 'not justified by our actual knowledge of gravitation' and was 'necessary only for the purpose of making possible a quasi-static distribution of matter, as required by the fact of the small velocities of the stars'. In other words, his cosmological constant was something of a fudge factor.

Now, however, other theoreticians entered the picture by exploring different relativistic models of the cosmos. A physicist and mathematician, Alexander Friedmann, proposed that non-static models of the universe should be considered. In 1922–24 he derived a whole class of dynamic cosmic models, using Einstein's field equations. Soon after, a theoretical physicist and Catholic priest, Georges Lemaître, treated emerging astronomical observations of a systematic recession of distant galaxies as evidence of an expansion of space on the largest scales. He postulated an origin of the universe in a 'primeval atom' – the first conception of what would become the Big Bang model that dominates cosmology today. Unaware of Friedmann's work, Lemaître demonstrated how such a cosmic expansion could be derived from Einstein's field equations. But when Einstein was made aware of the work of both Friedmann and Lemaître, he could not accept their cosmic models. In a discussion with Lemaître in 1927, Einstein said he regarded theoretical models of an expanding universe as 'totally abominable'.

In 1929, however, new observations compelled him to change this view. The American astronomer Edwin Hubble, working with the advanced telescope at the Mount Wilson Observatory in California and assisted by Milton Humason, observed a linear relation between the recession velocity of distant galaxies and their radial distance. 'Many theorists saw the phenomenon as possible evidence for an expansion of space, and set about constructing relativistic models of an expanding universe similar to those of Friedmann and Lemaître', according to a current physicist, Cormac O'Raifeartaigh. 'In all these theories, it was assumed that the average density of matter in the universe would decrease as space expanded – what is known as an "evolving" universe.'

Einstein own 'evolving universe' paper – a turning point in his thinking – was published in Berlin in April 1931, shortly before he departed for Oxford. His Oxford blackboard's final three mathematical symbols neatly summarize this cosmology paper, based on Friedman's relativistic model of an expanding cosmos, with the cosmological constant now set at zero and Hubble's measurements of the expanding universe used to estimate three quantities: the density of matter, ρ; the radius of the cosmos, P; and the timespan of the cosmic expansion, t (given as ten billion years). However, the blackboard's arithmetic was not totally accurate. 'It appears that Einstein stumbled in his use of the Hubble constant', notes O'Raifeartaigh, 'resulting in a density of matter that was too high by a factor of a hundred, a cosmic radius that was too low by a factor of ten, and a timespan for the expansion that was too high by a factor of ten'. No doubt this mistake was just one of the many 'untrue' elements that Einstein had in mind by the time he returned to Oxford in 1933 and gave the Herbert Spencer Lecture.

In this 1933 lecture, 'On the Method of Theoretical Physics', Einstein tried to escape from the messy physical reality inherent in experimental physics and astronomical observations, and substitute for it the paradise of pure mathematics, which he had been pursuing for some years in his unified field theory. 'The sanctuary from personal turmoil that Einstein sought in physics may also have coloured the extreme rationalist pronouncements in his Herbert Spencer Lecture', observes *The Cambridge Companion to Einstein*. Opinions consequently differ as to the lecture's value. One Einstein biographer, Albrecht Fölsing (who originally trained as a physicist), condemned it as 'a reckless overestimation of the possibility of understanding nature through mathematics alone – a mistake he would not have been capable of in his most productive years'. By contrast, another Einstein biographer, Abraham Pais, a practicing physicist who knew Einstein personally over a long period, hailed the lecture as 'perhaps the clearest and most revealing expression of his mode of thinking'.

It started with some typically Einsteinian humour. He ironically reprimanded himself for claiming more authority for his remarks to come than perhaps he should:

If you wish to learn from the theoretical physicist anything about the methods which he uses, I would give you the following piece of advice: don't listen to his words, examine his achievements. For to the discoverer in that field, the constructions of his imagination appear so necessary and so natural that he is apt to treat them not as the creations of his thoughts but as given realities.

This statement may seem to be designed to drive my audience away without more ado. For you will say to yourselves, 'The lecturer is himself a constructive physicist; on his own showing therefore he should leave the consideration of the structure of theoretical science to the epistemologist'.

Having openly voiced this subtle objection, Einstein further warned his listeners that his personal view of the past and present of theoretical physics was bound to be coloured by what he was currently trying to achieve and hoped to achieve in the future – an implicit reference to his ongoing work on the unified field theory (which he would briefly mention later in the lecture). 'But this is the common fate of all who have adopted the world of ideas as their dwelling-place. He is in just the same plight as the historian, who also, even though unconsciously, disposes events of the past around ideals that he has formed about human society.'

Then Einstein got to grips with the relationship in physics between pure theory, that is the free inventions of the human mind, and the data of experience, that is our observations of physical reality. 'We honour ancient Greece as the cradle of western science', he initially reassured his audience, many of whom were no doubt classically trained. 'She for the first time created the intellectual miracle of a logical system, the assertions of which followed one from another with such rigour that not one of the demonstrated propositions admitted of the slightest doubt – Euclid's geometry.'

Yet, for science to comprehend reality, more than Greek thought was required, he reminded his audience:

Pure logical thinking can give us no knowledge of the world of experience; all knowledge about reality begins with experience and terminates in it. Conclusions reached by purely rational processes are, so far as reality is concerned, entirely empty. It was because he recognized this, and especially because he impressed it upon the scientific world, that Galileo became the father of modern physics and in fact of the whole of modern natural science.

Isaac Newton, 'the first creator of a comprehensive and workable system of theoretical physics', agreed with Galileo's view, Einstein continued.

[He] still believed that the basic concepts and laws of his system could be derived from experience; his phrase '*hypotheses non fingo*' can only be interpreted in this sense. In fact, at the time it seemed that there was no problematical element in the concepts of space and time. The concepts of mass, acceleration and force, and the laws connecting them, appeared to be directly borrowed from experience. Once this basis is assumed, the expression for gravity seems to be derivable from experience; and the same derivability was to be anticipated for the other forces.

However, Newton was aware of certain difficulties:

One can see from the way he formulated his views that Newton felt by no means comfortable about the concept of absolute space, which embodied that of absolute rest; for he was alive to the fact that nothing in experience seemed to correspond to this latter concept. He also felt uneasy about the introduction of action at a distance. But the enormous practical success of his theory may well have prevented him and the physicists of the eighteenth and nineteenth centuries from recognizing the fictitious character of the principles of his system.

In other words, they did not recognize that the principles were free inventions of the human mind. 'On the contrary', Einstein explained, 'the scientists of those times were for the most part convinced that the basic concepts and laws of physics were not in a logical sense free inventions of the human mind, but rather that they were derivable by abstraction, i.e. by a logical process, from experiments'.

Then Einstein mentioned his own contribution:

It was the general theory of relativity which showed in a convincing manner the incorrectness of this view. For this theory revealed that it was possible for us, using basic principles far removed from those of Newton, to do justice to the entire range of the data of experience in a manner even more complete and satisfactory than was possible with Newton's principles. But quite apart from the question of comparative merits, the fictitious character of the principles is made quite obvious by the fact that it is possible to point to two essentially different principles, both of which correspond to experience to a large extent. This indicates that any attempt logically to derive the basic concepts and laws of mechanics from the ultimate data of experience is doomed to failure.

However, said Einstein,

If it is the case that the axiomatic basis of theoretical physics cannot be an inference from experience, but must be a free invention, have we any right to hope that we shall find the correct way? Still more – does this correct approach exist at all, save in our imagination? Have we any right to hope that experience will guide us aright, when there are theories (like classical [i.e. Newtonian] mechanics) which agree with experience to a very great extent, even without comprehending the subject in its depths?

He answered confidently, if controversially: 'in my opinion there is *the* correct path and, moreover, that it is in our power to find it'. Then followed the most quoted words in Einstein's Herbert Spencer Lecture (given below in italics):

> Our experience up to date justifies us in feeling sure that in nature is actualized the ideal of mathematical simplicity. It is my conviction that pure mathematical construction enables us to discover the concepts and the laws connecting them which give us the key to the understanding of the phenomena of nature. Experience can of course guide us in our choice of serviceable mathematical concepts; it cannot possibly be the source from which they are derived; experience of course remains the sole criterion of the serviceability of a mathematical construction for physics, but *the truly creative principle resides in mathematics.*
>
> In a certain sense, therefore, I hold it to be true that pure thought is competent to comprehend the real, as the ancients dreamed.

Thus, in his 1933 lecture, Einstein assured his Oxford audience – and by extension the international world of physics – that mathematics, on its own, could provide the basis for understanding nature. He apparently now rejected his own earlier position, as elegantly stated in 1921: 'As far as the laws of mathematics refer to reality, they are not certain; and as far as they are certain, they do not refer to reality'. As evidence of his new belief, he cited his own general theory of relativity, conceived in 1915–16, which he now claimed to have been essentially based on mathematical concepts rather than physical observations – for all its crucial confirmation by Eddington and fellow British astronomers in 1919. No doubt many theoretical and experimental physicists were taken aback and unconvinced by such a bold claim, even coming from Einstein, flying as it did in the face of so much of the history of physics, beginning with Galileo and Newton, that evidently arose from a combination of theory, observation and experiment. But which of them was in a position to contest its validity with general relativity's world-famous creator?

Perhaps they should have taken Einstein's initial humorous warning in Oxford more seriously. Was Einstein himself really the best authority on the origins of relativity theory? Maybe not. Several decades later, 'a team of scholars scrutinized his notebooks and demonstrated how he had developed his theory. They saw clear evidence that he used a two-pronged strategy, involving both mathematics and physical reasoning, right up until he completed the theory, yet he subsequently downplayed the role of physical reasoning', writes a current physicist and historian, Graham Farmelo, in *The Universe Speaks in Numbers*. 'Einstein appears to have largely based his new philosophy of research on distorted recollections of his work in the final month of his search for the correct field equations of gravity.' According to one of the team of scholars, Jeroen van Dongen, 'Einstein overemphasized the part mathematics had played in his development of his theory of gravity, probably to try to persuade his critical colleagues of the value of his way of trying to find a unified theory of gravity and electromagnetism'.

Ironically, there was a clear indication of Einstein's tendency towards such distorted recollection a mere ten days after his Herbert Spencer Lecture, in his lecture at the University of Glasgow on 20 June. Speaking in considerable technical detail on 'The Origin of the General Theory of Relativity', Einstein commented on the dark days of late 1915, after two years of excessively hard work, when he had felt lost. Finally, he recognized certain errors of thought, and 'after having ruefully returned to the Riemannian curvature, succeeded in linking the theory with the facts of astronomical experience'. In other words, general relativity rested not only on mathematical concepts such as Bernhard Riemann's non-Euclidean geometry of curved space, but also on physical – in this case astronomical –

observations. He closed the Glasgow lecture with the following memorable statement:

> In the light of the knowledge attained, the happy achievement seems almost a matter of course, and any intelligent student can grasp it without too much trouble. But the years of anxious searching in the dark, with their intense longing, their alternations of confidence and exhaustion and the final emergence into the light – only those who have experienced it can understand that.

Little wonder, then, that the origins of general relativity should be an obscure and contentious subject – even to Einstein himself – in Oxford in 1933.

Chapter 3
ARTISTIC INTERLUDES

Einstein's reflections on the origins of his ideas in physics, such as general relativity, are sometimes reminiscent of an artist rather than a scientist. As he himself remarked, 'The greatest scientists are artists as well'. Indeed, there are times when his science brings to mind the art of his undoubtedly favourite artist – Wolfgang Amadeus Mozart. Perhaps Einstein sensed this when he once said, 'Mozart's music is so pure and beautiful that I see it as a reflection of the inner beauty of the universe'.

His most evident artistic talents lay in music, playing mainly the violin from boyhood and the piano in later life, although he also showed skill in writing verse and also creating occasional caricatures – notably some paper cut-outs of himself and his family, of which he was justifiably proud. While in Oxford, he played a substantial amount of music with professional musicians, wrote a humorous poem, and also sat for a significant portrait. In addition, he attended a few public musical events, such as a concert of music by Joseph Haydn at the Sheldonian Theatre, a recital by an African-American singer, Roland Hayes, at the Town Hall, and a performance of old Coptic music at the University Church of St Mary the Virgin by an unnamed lecturer. This last was an extraordinary occasion. In his diary, Einstein described the lecturer standing in a blue habit on a blue podium as 'a fat giant with a red face' and 'a picturesque swindler'. More productively, he visited the

8a and b. The Deneke family and friends in the music room of Gunfield, their North Oxford house, 1930s – also depicted in a family Christmas card (opposite). Einstein played his violin there many times, and had a warm relationship with Margaret Deneke (seated at the piano), who vividly recorded his Oxford visits in her diary.

Ashmolean Museum with Lindemann to see the objects excavated from the mysterious Minoan civilization in Crete by Arthur Evans; they struck him as 'more Egyptian than Greek in character'.

His musical friend Margaret Deneke provided the most vivid vignettes of Einstein, among all of his contacts at Oxford, during his visit in 1931 and also his subsequent visits in 1932 and 1933 – recorded at length in her diary in translation from her frank conversations with Einstein. The younger of two surviving

daughters of a wealthy London merchant banker born in Germany
and his German-born wife (who was a close friend of the celebrated
pianist Clara Schumann), Margaret and her sister Helena were
born in London and later moved to Oxford. In 1913, Helena was
appointed bursar and tutor in German at Lady Margaret Hall,
the university's first women's college, while Margaret became a
musician/musicologist. Originally trained in the work of the great
German romantic composers – Ludwig van Beethoven, Johannes
Brahms, Franz Schubert and Robert Schumann – she then studied
modern English music under the influence of Oxford's musical
scholars, especially Ernest Walker and Donald Tovey, and was soon

interpreting English music to many school-children. Together the sisters lived in a Gothic villa named Gunfield, very close to Lady Margaret Hall, where they staged frequent performances in their music room by famous professionals, such as Adolf Busch and Marie Soldat-Roeger, as well as gifted amateurs. 'Generations of Oxford undergraduates, colleagues and friends enjoyed the Denekes' hospitality at the many Gunfield concerts', according to the *Oxford Dictionary of National Biography*'s entry on Helena Deneke. 'Marga Deneke, herself a talented pianist, was choirmaster at Lady Margaret Hall and, raising considerable sums of money through concerts and lecture recitals, became one of the college's benefactors.' The Oxford Chamber Music Society met at Gunfield free of charge for some twenty-seven years, with Margaret Deneke making up any financial deficits – especially during the Second World War, the Society owed its survival to the Deneke sisters' generosity. In May 1931, inevitably, they would be joined by a visiting amateur violinist: Einstein.

'In he came with short quick steps. He had a big head and a very lofty forehead, a pale face, a shock of grey, untidy hair', Margaret Deneke recalled of her first meeting with Einstein on 11 May. She introduced him to Marie Soldat, who was a violin virtuoso originally discovered by Brahms and a pupil of Joseph Joachim. After dinner, Einstein

turned to Mother with an engaging smile: 'You have provided a delightful meal; now the enjoyable part of our evening is over and we must get down to work on our instruments. Shall we play Mozart? Mozart is my first love – the supremest of the supreme – for playing the great Beethoven I must make something of an effort, but playing Mozart is the most marvellous experience in the world.'

The players started with Mozart; under Marie Soldat's rich tone on her Guarnerius del Gesu [a famous 1742 violin], the professor's borrowed violin sounded starved and raucous, but his rhythm was impeccable. Before passing on to Haydn, there was an interval for a chat. Marie Soldat commented on the professor's long violin fingers, tapering usefully at the tips. The professor said, 'Yes I never have practiced and my playing is that sort of playing, but physical build cannot be altogether divorced from mental gifts, an unusually sensitive temperament will make its mark on a body'.

I said Adolf Busch [an intimate friend of Einstein] had got short fat fingers.

'Well I must admit without any fingers at all no one can play a fiddle.'

Before they left, the guests signed the visitors' book. 'In his small clear handwriting', Deneke observed, Einstein wrote: 'Albert Einstein *peccavit*'. He 'sinned' again a few days later. The occasion was a somewhat strained formal dinner at Rhodes House, given in his honour by its warden, Francis Wylie, the evening before Einstein's second lecture on 16 May. (Pre-dinner, poor Einstein had had to grab a needle and thread in the bachelor sleeping quarters and sew up his ill-fitting dress shirt, so that his hairy chest did not peer out during dinner!) 'Lady Wylie thought he would enjoy himself more if music could be introduced', noted Deneke. 'Professor Einstein's English was rather halting in those days and to converse in French or German might not be too easy for the trustees.' When the meal was finished, he, Soldat and Deneke formed a trio. According to her diary:

We were established around the piano; the trustees saw their lion disporting himself with Bach's Double Concerto and Handel and Purcell Sonatas. He tucked the violin under his chin with a will and tuned long and loud. Then he chose the piece he wanted and made suggestions: 'No repeats in the Adagio please and the Allegro not too fast, there are tricky bits for me'. We started obediently and he threw himself whole-heartedly into the music. He made no effort at all to discover what the trustees might like to hear. Unashamedly we played for our own enjoyment, without the slightest pretence about performing. The trustees smoked in silence, witnessing their guest of honour having a happy evening. I doubt if they listened.

Indeed, according to Einstein's somewhat franker diary, when their music started up, 'The guests hastily left the room'!

On another occasion, Deneke's teacher, Tovey, came over to Gunfield for an evening of music. She, Soldat and another professional musician, cellist Arthur Williams, startled the group by playing Tovey's version of Haydn's Trio in A major with the parts redistributed for the benefit of the cello. 'Professor Einstein thought we had discovered a Mozart Trio that was unknown to him. He jerked up from the sofa and dashed to the piano to look at the music. How could there be a Mozart Trio that he had never come across?'

The following year, in May 1932, Einstein returned to Gunfield for further musical get-togethers. One night they finished the last note of Brahms's Piano Quintet at 11.20 p.m. Einstein called out: 'The gate at Christ Church is locked at eleven – what am I to do now?' Having rushed to the telephone and called the porters' lodge, they discovered the college would remain open until midnight (perhaps in Einstein's honour?). Before hurrying into the Denekes' car, Einstein scribbled a postcard to Tovey, thanking their musicologist friend for the gift of his book on the art of the fugue. 'Marie Soldat felt our evening had been merrier than any were last year', noted Deneke happily. 'The professor was more *sans gêne* and we others had overcome the shyness of having him as our guest.'

How good a musician was Einstein? Opinions have varied over the decades, in both Germany and beyond (not helped by the occasional confusion of Einstein with a distinguished musicologist, Alfred Einstein, his contemporary in both Germany and the United States). Clearly Deneke had some reservations about his playing, yet

9. Einstein as a violinist, possibly in 1933. He played violin on his three Oxford visits, and discussed music with Margaret Deneke and her fellow musicians – who admired some aspects of his playing.

she was well aware that Einstein always had to use a borrowed violin during his time in Oxford. When he returned in 1932, she recorded his advice about an instrument she was thinking of buying. 'He scrutinized the violin with great interest, and after he had played on it he advised me not to buy it as the tone was uneven.' Instead, he recommended she should acquire a cheap instrument and give it to the man who had treated his own violin by adding varnish and cutting away little bits to improve the tone. 'He had had no other violin than this treated cheap fiddle, thin in its wood, but clear in voice.' As for the opinion of professional musicians on Einstein, it is probably summarized by the comment, 'relatively good', from the pianist Artur Balsam, who once played with him.

Undoubtedly, Einstein was an intellectual match for professional musicians in the speed of mind essential for ensemble playing, even if he lacked their tone. At the same time, as a dedicated amateur quartet player, he was (to quote an American musicologist): 'the denizen of dimly lit music rooms of the world where enthusiastic friends and fiddlers bend together over their instruments and sleepy children yawn, and towards the end of a long evening someone says "Let's play some Haydn and then call it a night".'

By no means all of Deneke's Einstein diary for 1931–33 concerned music. She met him not only at Gunfield, but also in his rooms at Christ Church and elsewhere around Oxford, including Lady Margaret Hall, when he gave the packed Deneke Lecture on atomic theory in June 1933. She also persuaded him to sit for a portrait by a little-known Tyrolese peasant artist, F. Rizzi, who had come to Oxford not long before Einstein's arrival in May 1933 to do a portrait of Deneke's mother at the family's request. When Mrs Deneke unexpectedly died, Rizzi was left at a loose end, deprived of his main commission.

Rizzi had never heard of Einstein, but both he and his subject got on genially when Deneke brought Rizzi to Einstein's rooms in Christ Church. On 31 May she returned with two big bunches of flowers supplied by Lady Margaret Hall's gardens (which were designed and looked after for decades by her sister, Helena). She noted:

> Professor Einstein was still at his breakfast table and was delighted with the flowers. He had no coat on, no stockings but sandals on

the Porter of Ch Ch phoned 'Miss
Deneke's little man was wait-
= the Porch'. I bicycled down
to Ch Ch at once and found
Prof Rizzi surrounded by a group
of people who were admiring
his profile picture of Prof E.
looking down.

We took it to Soame the
photographer, and suddenly
Prof Rizzi remembered it was
not signed — So we returned
to Ch Ch. But Prof E. had
gone out for a walk —
the Ch Ch Meadow. We tried
to pursue him — the Meadow
Porter had noticed exactly
where he had gone, but
he was too far ahead. So
we returned to Gunfield, leaving

10. Page from Margaret Deneke's Oxford diary, June 1933. It captures her amusing
experience of arranging for the artist F. Rizzi to draw a portrait of Einstein in his rooms
in Christ Church.

his bare feet. He pointed to the pile of letters and complained that correspondence was a great burden. I confirmed June 13 for the Deneke Lecture and at his request made this entry in his diary. Then I asked if he would let Rizzi sketch him – we wanted work for the painter. He consented: 'he can draw my portrait whilst I am at work; then I sit still anyhow.' I promised Rizzi would be silent.

The following morning, she brought the artist again, installed him on some curious steps leading to the room's main window overlooking the college's famous cathedral, and left him alone with Einstein. Around noon, the Christ Church porter – who had been instructed to protect Professor Einstein from unscheduled visitors – phoned Gunfield to say that 'Miss Deneke's little man was waiting in the porch'. Immediately bicycling to Christ Church, she found Rizzi in the porch surrounded by a group of people who were admiring his profile portrait of Einstein looking down at a book. (Today it hangs in Christ Church's Senior Common Room.) The two of them then carried it off to a photographer, but all of a sudden Rizzi remembered that 'Professor Eisenstein', as he habitually called Einstein, had not signed the portrait. They returned to Christ Church, and found that their quarry had gone for a walk in Christ Church's meadows. The two of them set off in hot pursuit, but Einstein had got too far ahead, and so the signature had to wait until the artist returned the next day.

When Einstein finally signed the portrait in his college rooms, Rizzi told Deneke that he laughed heartily over the artist's German aphorism: '*Nichts koennen ist noch lange keine Moderne Kunst*' ('To be a Modern artist, it is not enough to lack any skills'). Einstein's taste in drawing, like his taste in music, was always firmly in favour of the classical.

11. Drawing of Einstein by F. Rizzi made in Christ Church, Oxford, June 1933. It is now kept on display in the college's Senior Common Room.

Chapter 4
STUDENT OF CHRIST CHURCH AND OXFORD

On one of several visits to Einstein in Christ Church, Deneke found herself among undergraduates streaming out of a lecture through the grandeur of Tom Quad and passing close to the two of them. 'I was shocked to see how much they stared at Professor Einstein, and I was amused to see how little he seemed to notice it.' Yet, even if Einstein was not much aware of his impact on Oxford students, he proved to be an observant, and sometimes amusing, student of Oxford life.

Appropriately enough, Einstein's first set of Christ Church rooms, in 1931, overlooking Tom Quad, had been occupied in the later nineteenth century by the mathematician (and clergyman) Charles Lutwidge Dodgson – best known as Lewis Carroll, the author of *Alice's Adventures in Wonderland.* Carroll's poem 'The Walrus and the Carpenter' inspired a well-known parody, 'The Einstein and the Eddington', a nonsense poem written in 1924 by an American professor specializing in relativity. By 1931 these rooms had become home to a war veteran and classical scholar, Robert Hamilton Dundas, who, luckily for Einstein, was away from England in India as part of a world tour in 1930–31. When Dundas returned, he was charmed to find that Einstein had written a poem of his own in the visitors' book, inspired by the rooms with bookshelves full of classical literature and – who knows? – the lingering spirit of Dodgson. More doggerel than Carroll, the poem is nonetheless thoughtful and witty, in this free translation from Einstein's rhymed German by an Oxford literary scholar, J.B. Leishman:

Dundas lets his rooms decay
While he lingers far away,
Drinking wisdom at the source
Where the sun begins its course.

That his walls may not grow cold
He's installed a hermit old,
One who undeterredly preaches
What the art of Numbers teaches.

Shelves of towering folios
Meditate in solemn rows;
Find it strange that one can dwell
Here without their aid so well.

Grumble: Why's this creature staying
With his pipe and piano playing?
Why should this barbarian roam?
Could he not have stopped at home?

Often, though, his thoughts will stray
To the owner far away,
Hoping one day face to face
To behold him in this place.

A year later, Einstein and Dundas – the 'barbarian' and his bibliophile host – did meet, when Einstein returned to Christ Church in 1932. The poem was published in *The Times* soon after Einstein's death, and in due course the visitors' book was donated to the Bodleian Library.

Einstein's joke against himself refers to perhaps his only serious reservation about life in Christ Church (and indeed the

Dundas will im heil'gen Osten
Seine Bude kommt' verrosten
Während er nach Hohem strebt
Und in weiter Ferne lebt.

Dass die Mauern nicht erkalten
Nimmt er hier herein den Alten
Der da predigt unentwegt
Und die Rechenkünste pflegt.

Der Folianten ernste Reih
Denket sich gar mancherlei
Wundern sich, dass so ein Mann
Ohne sie hier hausen kann.

Grollen: Was will dieser hier?
Raucht Tabak und spielt Klavier
Der Barbar geht ein und aus
Wishalb blieb er nicht zuhaus?

Dieser aber oft und gerne
Denkt des Hausherrn in der Ferne
Freut sich, bis er diesen Mann
In der Nähe sehen kann.

Mit herzlichem Dank und Gruss

Albert Einstein

Mai 1931.

12. Visitors' book from Einstein's rooms in Christ Church, Oxford, May 1931, now in the Bodleian Library. His German poem thanks his college host Robert Dundas; its translation (opposite) appeared in *The Times* after Einstein's death in 1955.

EINSTEIN BY NUMBERS

AN OXFORD MANUSCRIPT

While Einstein was a Research Student at Christ Church he used, in 1931, the rooms of Mr. R. H. Dundas, who was, at that time, travelling round the world. On his return Mr. Dundas found the verses given below in his Visitors' Book in Einstein's handwriting.

Dundas weilt im heil'gen Osten
Seine Bude könnt' verrosten
Während er nach Hohem strebt
Und in weiter Ferne lebt.

Dass die Mauern nicht erkalten
Nimmt er hier herein den Alten
Der da predigt unentwegt
Und die Rechenkünste pflegt.

Der Folianten ernste Reih
Denket sich gar mancherlei
Wundern sich, dass so ein Mann
Ohne sie hier hausen kann.

Grollen: Was will dieser hier ?
Raucht Tabak und spielt Klavier
Der Barbar geht ein und aus
Weshalb blieb er nicht zuhaus ?

Dieser aber oft and gerne
Denkt des Hausherrn in der Ferne
Freut sich, bis er diesen Mann
In der Nähe sehen kann.

Mit herzlichem Dank und Gruss
   Albert Einstein
   Mai 1931

This free English version is by Mr. J. B. Leishman, of St. John's College: —

Dundas lets his rooms decay
While he lingers far away,
Drinking wisdom at the source
Where the sun begins its course.

That his walls may not grow cold
He's installed a hermit old,
One that undeterredly preaches
What the Art of Numbers teaches.

Shelves of towering folios
Meditate in solemn rows;
Find it strange that one can dwell
Here without their aid so well.

Grumble: Why's this creature staying
With his pipe and piano-playing ?
Why should this barbarian roam ?
Could he not have stopped at home ?

Often, though, his thoughts will stray
To the owner far away,
Hoping one day face to face
To behold him in this place.

With hearty thanks and greetings.

"The Barbarian" and his host did meet when the latter came home to Christ Church.

university as a whole): its penchant for formality, symbolized by the dreaded dinner-jacket. 'He was by nature a rebel who enjoyed being unconventional', wrote his later scientific collaborator in Princeton, Oxford-educated Banesh Hoffmann. 'Whenever possible he dressed for comfort, not for looks.' For Einstein, such formalities were reminiscent of Prussian *Zwang* ('compulsion' or 'coercion') that he resisted from childhood, through his school and college education in Germany and Switzerland, his years of fame in Berlin and the rise of Nazism, right up to his death in Princeton.

His earliest two diary entries after arrival in the college with Lindemann in May 1931 catch the flavour of this reservation well. 'Evening club meal in dinner-jacket. The apartment is reminiscent of a small fortress and belongs to a philologist who is currently in India. Young servant; communication droll', he noted on 1 May. 'The dinner takes place with about 500 professors and students, the former all in dinner-jackets and black gowns, a bizarre as well as boring affair. Of course the service is men only. One gets a slight idea of how horrible life would be without women. All in a kind of art basilica!' (Today Christ Church Hall displays Einstein in a stained-glass window.) And in his next, briefer entry on 2/3 May he wrote: 'Silent existence in the hermitage in bitter cold. In the evening the solemn Communion Supper of the holy brotherhood in tails. On 3 May with deans (clergymen), who introduce me as a quasi-guest, taciturn and solemn but benevolent with delicate jokes on the tips of their tongues.' Among them was of course Dean White, who would later fall asleep at Einstein's third, and toughest, Rhodes lecture. Although, over time, Einstein came to appreciate Christ Church's English reserve somewhat better, he would never have much time for its clerical aura.

13. Einstein depicted in a stained-glass window of today's Christ Church Hall, Oxford. After dining there many times in the early 1930s with mixed emotions, he described the Hall as 'a kind of art basilica!'

Lindemann, and some other Christ Church colleagues – notably economist Roy Harrod and classicist Gilbert Murray – compensated for the overall atmosphere of formality. Einstein 'was a charming person, and we entered into relations of easy intimacy with him', Harrod recalled in his memoir of Lindemann. (Like Lindemann, however, Harrod thought Einstein 'naïve' in human affairs.) Einstein apparently divided his time between playing his violin and his mathematics. While crossing Tom Quad, 'one was privileged to hear the strains coming from his rooms'. And in the college's governing-body meetings, 'I sat next to him; we had a green baize table-cloth; under cover of this he held a wad of paper on his knee, and I observed that all through our meetings his pencil was in incessant progress, covering sheet after sheet with equations.' Murray, who had befriended Einstein while they were serving together on the League of Nations Committee on Intellectual Cooperation in the 1920s, remembered coming across Einstein in 1933 – now a refugee from Nazi Germany – sitting alone in Tom Quad with a smiling, faraway look on his face. 'Dr Einstein, do tell me what you thinking.' Einstein replied: 'We must remember that this is a very small star, and probably some of the larger and more important stars may be very virtuous and happy'.

Even before Einstein left Oxford for Berlin on 28 May 1931, Lindemann appears to have begun negotiations within Christ Church to lure him back. His idea was that Einstein should be elected a 'research student' (namely, a fellow) of the college and be offered a bursary from college funds so that he could spend relatively brief periods of time in Oxford each year at his own convenience. By late June, Dean White informed Lindemann that the college was in a position to offer a 'studentship' to Einstein for five years – with an annual stipend of £400, a dining allowance and accommodation during his periods of residence, which it was hoped would last for about a month each year during Oxford term time. Lindemann promptly intimated this possibility to Einstein in a letter.

Einstein was immediately tempted – not least because of the disastrous economic and political situation at home in Germany, with banks at risk of collapse, escalating unemployment and the rising popular appeal of the Nazi Party. 'Your kind letter has filled me with great pleasure and brought back to me the memory of the wonderful weeks in Oxford', he wrote to Lindemann on 6 July. By mid-July, it became clear that he was ready to accept Christ Church's offer. But in a letter to the dean Einstein did not clarify whether he accepted all of the terms of the studentship, in particular its annual residence requirement. Lindemann persuaded the college to give Einstein the benefit of the doubt: although he might spend less than a month at Christ Church in any one year, he would make up for this during another year. On 21 October, the governing body elected Einstein unanimously to a research studentship. Einstein accepted the offer on 29 October with a warm letter to the dean referring to the college's 'harmonious community life'.

In the interval, however, White received a stern letter of protest at the college's decision. Dated 24 October, it came from John George Clark Anderson, a former tutor and student of Christ Church who had taught classical history there in 1900–27, before migrating to Brasenose College when he became Camden Professor of Ancient History. He wrote:

> I cannot help thinking that it is unfortunate that an Oxford college should send money out of the country in the present financial situation and at a time when the university is receiving a large government grant at the expense of the taxpayers for educational purposes. The more I think of it, the more strange does this new development appear to me ...

At no point in Anderson's letter did he mention Einstein by name or give even the merest hint that he was aware of the scientist's eminence. The dean took up this point in his reply:

I think that in electing Einstein we are securing for our society perhaps the greatest authority in the world on physical science; his attainments and reputation are so high that they transcend national boundaries, and any university in the world ought to be proud of having him. Then in spite of his scientific position he is a poor man, and this quite moderate pecuniary help will enable him to carry on his work better. ... It is quite true that we are paying money out of England; but I think we are getting more than our money's worth.

A further letter from Anderson expanded at length on his arguments about funding, some of which were reasonable, especially in a time of serious economic depression. Yet, the emotions underlying the arguments were clear from two comments: that a research studentship should not go to 'a German who has no connection with the university (beyond being the recipient of an honorary degree)'; and that 'it does not seem to me to be patriotic, especially in such times as the present, to use college revenues to endow foreigners. Charity should always begin at home'. Anti-German feeling and Oxford academic parochialism were still, sadly, alive and well in 1931. To the credit of Christ Church, its governing body rose above such attitudes with Einstein, well before he became a target of Nazi attacks in 1933.

Amusingly, even though the British tax authorities agreed with Anderson's argument about money going out of England, they did not accept the negative conclusion he drew from it. In 1932, a zealous Oxford tax inspector suggested that Einstein's Christ Church stipend should be liable for income tax. Lindemann, concerned, discussed the matter with the chairman of the board of the Inland Revenue and the Revenue's special expert on the subject. According to the latter, Einstein's annual stipend from Christ Church should be exempt from British tax. Einstein did not live or work in Oxford, he wrote, and thus, 'The world at large, and not Christ Church in particular, gets the benefit of his work'. Although England would get its money's worth from Einstein's visits to Oxford, his true value would have no national boundaries.

Einstein's relationship with Christ Church in 1931–33 proved amiable but eccentric from beginning to end. But on the whole – at least judging from his Oxford diary entries – his most fruitful contacts took place less in the college than in other settings (as understood by the tax expert). These ranged from other colleges and university societies to private houses and more informal meetings, including, of course, the dinner-concerts at Gunfield organized by Deneke.

Next to physics and music, Einstein's other preoccupation in Oxford was political, on each of his three visits. For example, on 22 May 1931 he visited Ruskin College, an independent institution set up as Ruskin Hall in 1899 with the support of the trades union movement to offer courses for working-class students unable to gain access to the University of Oxford. Ruskin students were permitted to attend the university's lectures. Having met its forty or so young men and women, Einstein pronounced Ruskin an 'excellent institution', and noted that nineteen British Members of Parliament were former members of Ruskin.

He also had meetings with political activists from both inside and outside the university. On 20 May, for example, he attended the League of Nations Society, and answered questions for two hours on internal German and Russian affairs. And on 23 May, on the evening of the day of his doctoral ceremony, he met pacifist university students at a private house, and was impressed by their political maturity – especially as compared with their 'pitiful' German equivalents in Berlin.

Another meeting, on 26 May, with members of War Resisters' International under its chairman, the British Member of Parliament A. Fenner Brockway – who had served time in prison during the First World War for his anti-conscription activities – brought forth a categorical response from Einstein. As Brockway recorded, Einstein exclaimed:

> There are so many fictitious peace societies. They are prepared to speak of peace in time of peace, but they are not dependable in time of war. Advocates of peace who are not prepared to stand for peace in time of war are useless. To advocate peace and then to flinch when the test comes means nothing – absolutely nothing.

Surely this was a statement Einstein would come to regret by August 1933, when he publicly abandoned pacifism after the Nazi rise to power.

A final meeting, with the Quaker-run Friends' Peace Committee on 27 May 1931, was probably the most revealing of all about Einstein's pacifist views. A report of it in *The Friend* opened with a reference to his recent controversial 'Two-per-cent speech' in the United States, in which Einstein had argued that: 'Even if only two per cent of those assigned to perform military service should announce their refusal to fight, as well as urge means other than war of settling international disputes, governments would be powerless, they would not dare send such a large number of people to jail'. The report then discussed his views of how to organize war resistance in countries without conscription (such as post-war Britain), and also how to promote peace internationally. 'For example, he suggested that it is very important to organize the churches and get declarations from them against participation in war', in particular the Roman Catholic Church, 'as this would have so much influence in France and Italy'. As for diminishing Germany's rising militancy, Einstein advised that pacifist endeavours to denounce the war guilt clause in the Versailles peace treaty were less likely to work than international efforts to relieve Germany's 'terrible economic depression' by finding a solution to its war reparations problem. But he was not optimistic when asked about the prevention of war by international diplomacy, no doubt because of his experiences with the International Committee on Intellectual Cooperation. 'He pointed out that the machinery is there in the League of Nations, but that unfortunately in his view it is too weak. He believes, however, that the machinery is capable of being made strong and effective. "That depends,' he said, 'on what the people *will*".'

Apropos, Einstein paid an unexpected tribute to religion:

> To Friends, perhaps the most interesting part of the interview was Professor Einstein's statement that his attitude to war was held because he can 'do no other'. He agreed that the strongest anti-war convictions are those with a religious basis. People may be convinced intellectually that war is futile, but that is not enough. This greatest of thinkers does not trust to the intellect alone. 'Reason,' he says, 'is a factor of secondary importance'.

Yet it would be untrue to leave the impression that all of Einstein's time in Oxford during May 1931 was simply taken up with scientific, musical and political activities. As his diary made plain, he had a varied range of other experiences.

For example, Douglas Roaf, a physics research student at Christ Church, heard of Einstein's love of sailing and took him out on the River Thames in a skiff down to the Abingdon Cut. Seeing that Roaf was properly attired for the occasion and wearing plimsolls, Einstein offered to take off his brown boots. On another occasion, he attended a rowing regatta by the river. And he saw at first hand the sporty side of Lindemann, who had been a championship tennis player in both Germany and England in his younger days. While Lindemann played squash in a private squash court, Einstein watched from the spectators' gallery, as the mysteries of the game were explained to him by one of his undergraduate friends at Christ Church, Alfred Ubbelohde (later professor of thermodynamics at Imperial College in London).

14. Einstein on one of his many walks in central Oxford, early 1930s. The precise location is unknown. Once he happened to meet the undergraduate William Golding, who vividly described their encounter.

He also enjoyed plenty of informally dressed walking in Oxford, either with others, such as Deneke and Hermann Fiedler, the professor of German language and literature, or, quite frequently, wandering on his own in the streets and parks. As he told Deneke (who tried to cajole the somewhat dishevelled Einstein into dressing less like a dreamy professor, if without much success):

'The types of humanity in the streets here are interesting to me – they are quite unlike those seen at home' – that is, in Berlin.

In addition, he was taken out of Oxford by car into the Cotswold countryside, the beauty of which he much appreciated. Lindemann introduced him to his 85-year-old father at his country house in the Thames valley, who impressed Einstein with his liveliness. Adolph Lindemann, besides being an amateur astronomer keenly interested in relativity, had been an engineer and businessman making ships for the Russian navy under the tsars. He spoke of corruption in Russia. Having delivered ten ships, he was paid for nine of them. When he asked the government official in charge of the purchase about payment for the tenth ship, the official replied: 'Make sure you get home as soon as possible'. At another country house, belonging to a friend of Lindemann's, old Lady Fitzgerald, Einstein was tickled to see a strange, new, technological advance: drinking fountains for cows. When a cow pressed her muzzle against a metal plate, a valve opened and the water flowed into a drinking bowl. 'Soon there will be water-closets for cows', mused Einstein, the former patent clerk, in his diary. 'Long live civilization!'

Not recorded in the diary was his brief flirtation with a woman who had followed him from Germany to Oxford. In total contrast to the musical Margaret Deneke, Ethel Michanowski was a Berlin socialite. Einstein composed a brief poem for her on a notecard from Christ Church, which began: 'Long-branched and delicately strung,/ Nothing that will escape her gaze'. Michanowski then sent him an expensive present, which he did not welcome. 'The small package really angered me', he wrote to her. 'You have to stop sending me presents incessantly… And to send something like that to an English college where we are surrounded by senseless affluence anyway!'

But probably the most evocative memory of Einstein in Oxford concerned simply his charisma. It was recalled in an account of a chance encounter by William Golding, the future author of *Lord of the Flies* and Nobel laureate, who started as an undergraduate in science and then changed to literature. Some time in 1931, the young Golding happened to be standing on a small bridge in Magdalen Deer Park looking at the river when a 'tiny moustached and hatted figure' joined him. 'Professor Einstein knew no English at that time, and I knew only two words of German. I beamed at him, trying wordlessly to convey by my bearing all the affection and respect that the English felt for him.' For about five minutes the two stood side by side. At last, said Golding, 'With true greatness, Professor Einstein realized that any contact was better than none.' He pointed to a trout wavering in midstream. '*Fisch*', he said. Golding continues:

> Desperately I sought for some sign by which I might convey that I, too, revered pure reason. I nodded vehemently. In a brilliant flash I used up half my German vocabulary: '*Fisch. Ja. Ja.*' I would have given my Greek and Latin and French and a good slice of my English for enough German to communicate. But we were divided; he was as inscrutable as my headmaster.

For another five minutes, the unknown undergraduate Englishman and the world-famous German scientist stood together. 'Then Professor Einstein, his whole figure still conveying goodwill and amiability, drifted away out of sight.'

Chapter 5
REFUGEE FROM NAZISM

When Einstein's acceptance of the 1931 Rhodes lectures in Oxford was first announced in December 1930, there was an ominous comment from the *Jewish Telegraphic Agency*: 'The movement to induce Prof. Einstein to settle permanently in England after his summer stay in England has gained momentum here as a result of a recent report from Berlin to the effect that Einstein may not return to Germany in case the Hitlerites obtain control of that country.' This was perhaps the first clear portent of political events that would overshadow Einstein's relationship with Oxford and Britain as a whole during 1933.

The clouds really began to darken in May 1932, after Einstein returned from his second month in Oxford. While he was away a secret deal had been struck on 8 May between an influential military leader, General Kurt von Schleicher, and Adolf Hitler, head of the Nazi Party. This was followed on 30 May by the official removal of the chancellor, Heinrich Brüning, and the appointment of a new, ultra-right-wing chancellor, Franz von Papen, with Schleicher as his minister of defence – all at the behest of the aged president, Field Marshal Paul von Hindenburg. On 2 June, a German Christian leader and Nazi deputy, Wilhelm Kube, announced to the Prussian Diet that 'when we clean house the exodus of the Children of Israel will be a child's game in comparison', adding that 'a people that possesses a Kant will not permit an Einstein to be tacked onto it'. Thus began the slide towards the Nazi seizure of power in January–February 1933 and, soon after, the escape of Einstein and his wife Elsa from their native country to a temporary home in Belgium, which would propel him back to Oxford for his final stay in May–June 1933.

The idea started with a brief letter on 1 May 1933 from Einstein in Belgium to Lindemann in Oxford. Could he visit Oxford again in June? Might Christ Church be able to offer him a small room? It need not be so 'grand' as his previous rooms in 1931 and 1932. Then Einstein abruptly switched subjects and alluded to politics: 'You have probably heard of my little duel with the Prussian Academy' – from which he had resigned in late March. 'I shall never see the land of my birth again.' Finally he mentioned physics: he had just worked out with his calculator, Walther Mayer, 'a couple of wonderful new results of a mathematical-physical kind'.

Lindemann replied immediately on 4 May, saying that he would have written earlier but had had no address for Einstein in Belgium, and that he gathered from the newspapers that 'there was not much prospect of a letter to Berlin being forwarded'. He naturally welcomed Einstein's proposed visit, but suggested that, since Oxford's Trinity term would end on 16 June, Einstein should arrive at the end of May. A set of rooms would be available, but 'as we did not know your plans I am afraid they will be somewhat smaller than last year'.

Then Lindemann discussed his own recent visit to Berlin, for four or five days in mid-April over Easter, when he had seen many of Einstein's German colleagues. About the Prussian Academy's condemnation of Einstein and the wider political situation, he commented:

The general feeling was much against the action taken by the Academy, which was the responsibility of one of the secretaries without consultation with the members. I can tell you more about

Der Hausknecht der Deutschen Gesandtschaft in Brüssel wurde beauftragt, einen dort herumlungernden Asiaten von der Wahnvorstellung, er sei ein Preuße, zu heilen.

15. Nazi caricature of Einstein, April 1933. It shows him being booted out of the German embassy in Brussels, as a Jew, while living in Belgium. This was shortly before his final visit to Oxford.

it when you come. Everybody sent you their kind regards, more especially Schrödinger, but it was felt that it would be damaging to all concerned to write to you, especially as the letter would almost certainly not be forwarded. Conditions there were extremely curious. It seems, however, that the Nazis have got their hands on the machine and they will probably be there for a long time.

Lindemann went on to consider what could be done to help some of the threatened German–Jewish physicists, by trying to find them positions at Oxford. 'I need scarcely say that very little money is available and that it would cause a lot of feeling, even if it were possible to place them in positions normally occupied by Englishmen.' He specified two promising individuals, Hans Bethe and Fritz London (the first of whom would win a Nobel prize), recommended to him by Einstein's non-Jewish colleague, the theoretical physicist Arnold Sommerfeld. Lindemann asked whether Einstein would be willing to recommend Bethe and London too, while adding characteristically:

Perhaps there are others whom you might consider better, but I have the impression that anyone trained by Sommerfeld is the sort of man who can work out a problem and get an answer, which is what we really need in Oxford rather than the more abstract type who would spend his time disputing with the philosophers.

This was the launch of Lindemann's historic campaign to obtain British commercial funding for notable refugee experimental physicists to come to Oxford in 1933–34. He would help a group of distinguished Jews, including the low-temperature physicist Franz (later Sir Francis) Simon – who became a professor at Oxford, worked on the atomic bomb project, and eventually took over from Lindemann as head of the Clarendon Laboratory.

Einstein responded on 7 May, advising that he would try to reach England on 21 May. Lindemann should not go to any bother about his accommodation, because he needed only 'one room' in order to be comfortable. About politics, he added ominously: 'I think that the Nazis have got the whip hand in Berlin'. He had been informed on good authority that they were hurriedly collecting war materiel, notably aeroplanes. 'If they are given another year or two the world will have another fine experience at the hands of the Germans.' As for the two physicists mentioned by Lindemann, he recommended London as 'a great source of strength' but said he knew too little about Bethe to express an opinion. Regardless of which individuals might be chosen, he was very grateful to Lindemann for his efforts to relieve refugee physicists. He offered to give a third of his salary that year to help his threatened German–Jewish colleagues.

Then on 9 May, Einstein wrote again very briefly, announcing that he would not be able to arrive in Oxford until about 26 May, because his younger son had been taken seriously ill. He could not wait six weeks before seeing him in Zurich, if he were to have any mental peace in England. 'You are not a father yourself, but I know you will certainly understand.' Not long before heading for Oxford in late May, he would see Eduard, a long-term psychiatric patient suffering from schizophrenia, who was being looked after by professional clinics in Zurich. Although Einstein would remain in touch with his son by letter, he would never meet him again. (Eduard Einstein survived his father, dying in a Swiss psychiatric hospital in 1965.)

One decidedly puzzling omission from this May correspondence with Lindemann was the Herbert Spencer Lecture in Oxford, which Einstein had agreed to give in Trinity term 1933 while he was visiting Oxford in May 1932. Neither Einstein nor Lindemann referred to the lecture. Although it actually took place at Rhodes House on 10 June 1933 (as discussed in Chapter 2), it would appear that this date must have been settled by Einstein at the very last minute, relatively speaking – either shortly before he left Belgium in mid-May or perhaps soon after he arrived in Oxford on 26 May. Whichever was the case, the lecture must surely have been prepared under conditions of much personal and professional stress for Einstein, exacerbated by the Nazi announcement in mid-May of the seizure of his and Elsa's German financial assets. (No doubt the lecture fee, agreed as £525 in 1932, was now even more welcome

to him, especially when compared to the fee for the subsequent Deneke Lecture on 13 June, a mere £26.)

As he informed his friend Max Born – a physicist who had just escaped from Germany to the Italian Alps and would soon settle in Britain – on 30 May, in a letter sent from the cloistered calm of Christ Church: 'You know, I think, that I have never had a particularly favourable opinion of the Germans (morally and politically speaking). But I must confess that the degree of their brutality and cowardice came as something of a surprise to me'. He concluded: 'I've been promoted to an "evil monster" in Germany, and all my money has been taken away from me. But I console myself with the thought that the latter would soon be gone anyway'.

At a public event in Oxford's imposing University Museum on 2 June, Einstein appeared sorely in need of public reassurance. He had been invited to offer a vote of thanks for a lecture by Ernest Rutherford to the Junior Scientific Society on 'The Artificial Transmutation of the Elements'. Not only was Rutherford a Nobel laureate, like Einstein, he was also a peer of the realm, the 1st Baron Rutherford, and, in addition to his honours, a big booming extrovert with a voice loud enough to disturb sensitive scientific experimental apparatus – according to a standing joke among his Cavendish Laboratory co-workers in Cambridge. Lord Rutherford made quite a contrast with the smaller figure of Einstein, a theoretician who typically worked alone at home or in an office, never in a laboratory, and was naturally unconfident at public speaking in English. According to an Oxford undergraduate at the lecture, C.H. Arnold, Einstein seemed 'a poor forlorn little figure' beside Rutherford. While Einstein was delivering his speech of thanks, somehow coping with English, 'it seemed to me that he was more than a little doubtful about the way in which he would be received in a British

16. Einstein in the main quadrangle of Christ Church, Oxford, June 1933. At this point he was a refugee from Nazi Germany, living in Belgium, before settling in the United States.

university'. However, the moment he sat down, Einstein was greeted by a thunderous outburst of applause. As Arnold vividly recalled more than three decades later:

Never in all my life shall I forget the wonderful change which took place in Einstein's face at that moment. The light came back into his eyes, and his whole face seemed transfigured with joy and delight when it came home to him in this way that, no matter how badly he had been treated by the Nazis, both he himself and his undoubted genius were at any rate greatly appreciated at Oxford.

No wonder that his own lecture at Rhodes House on 10 June opened with an uncharacteristically emotional introduction – at least by Einstein's unemotional standards – perhaps hinting at his troubled mental state:

I wish to preface what I have to say by expressing to you the great gratitude which I feel to the University of Oxford for having given me the honour and privilege of delivering the Herbert Spencer Lecture. May I say that the invitation makes me feel that the links between this university and myself are becoming progressively stronger?

But his hope was not to be realized. After delivering the lecture, followed by the Deneke Lecture on 13 June, Einstein left Oxford for Glasgow, and soon returned to Belgium. He would never see Oxford again.

He did return to England, however. As described in my book *Einstein on the Run: How Britain Saved the World's Greatest Scientist*, in late July 1933 he met various British politicians, including Winston Churchill, to discuss the Nazi threat to world peace and to German Jews. In mid-September, now under personal threat of assassination by Nazi extremists in Belgium, at the insistence of his deeply worried wife he settled alone in a secret English rural location in Norfolk, near Cromer, guarded by a Member of Parliament and veteran of the First World War, Commander Oliver Locker-Lampson.

From here he returned to London for a few days in early October and addressed a public meeting – along with leading British figures from the worlds of science, politics, education and religion – at the Royal Albert Hall, which was intended to raise charitable donations for academic Jewish refugees from Germany. Einstein, as the star attraction, spoke on 'Science and Civilization' in his hesitant, peculiar and touching English, to massive applause from an overflowing audience (which included Blackshirts from the British Union of Fascists). Without the 'intellectual and individual freedom' won by our ancestors, said Einstein, 'there would have been no Shakespeare, no Goethe, no Newton, no Faraday, no Pasteur and no Lister' – and of course no Einstein. As William Beveridge, another speaker, remarked in his live broadcast on BBC radio that evening: 'I had never seen him before. Einstein was a legend to me. It is like seeing Christopher Columbus or Julius Caesar.' Afterwards, on the steps of the hall, Einstein told a newspaper reporter: 'I could not believe that it was possible that such spontaneous affection could be extended to one who is a wanderer on the face of the earth. The kindness of your people has touched my heart so deeply that I cannot find words to express in English what I feel.' He concluded: 'I shall leave England for America at the end of the week, but no matter how long I live I shall never forget the kindness which I have received from the people of England'.

On 7 October 1933, Einstein – now joined by Elsa – boarded a liner in Southampton, heading for New York. Never returning to Europe, he died in Princeton, New Jersey, in 1955.

17. Einstein (front row, second from left) as a guest at the Oxford Union, June 1933. He did not speak in its political debate, but held many discussions about politics elsewhere in Oxford.

Postscript

EVERYWHERE AND NOWHERE AT HOME

Einstein undoubtedly enjoyed himself in Oxford, judging from his 1931 diary and the diary of Margaret Deneke in 1931–33 – unlike his fellow physics Nobel laureate Erwin Schrödinger, who spent a distinctly unhappy three years in Oxford in 1933–36 as a fellow of Magdalen College (as described in *Schrödinger in Oxford* by David Clary, a recent president of Magdalen). Einstein was also undoubtedly grateful to Oxford as a place of refuge from Nazi Germany in 1933. But what did the city and its university mean to him in the longer term? This question is not easy to answer, because Einstein scarcely commented on Oxford during his years in America, even in his 1933–55 correspondence with his friend Max Born, who liked Oxford, and in his conversations with notable, academically minded visitors from Britain, such as scientist and writer C.P. Snow, writer and literary critic V.S. Pritchett, and Margaret Deneke.

Perhaps the most revealing clue lies in his thank-you poem now at the Bodleian Library (quoted in Chapter 4), penned near the end of his first Oxford visit in 1931. It is humorous, evidently – but it is also serious, if not quite so evidently. In its five verses, its peripatetic author pictures himself as both a 'hermit' and a 'barbarian', dedicated above all else to the solitary pursuit of mathematics, 'the art of Numbers', and indifferent to books, 'the shelves of towering folios'. This is a recognizable portrait of Einstein as a young physicist in Switzerland around 1900, Einstein as a researcher on general relativity in Germany in 1915, and Einstein as an older figure in America in the 1940s and '50s obsessed with his unified field theory. In other words, a man who was both everywhere and nowhere at home: a citizen of the world, yes, but also 'a wanderer on the face of the earth' (to recall his own heart-felt words to the London journalist just before he left Britain in October 1933) and even 'an old gypsy' (as he wrote to a long-standing friend in Switzerland from Princeton in 1950). Home, for Einstein, was always his own mind.

The distinguished Oxford-based philosopher Isaiah Berlin, who met Einstein in America in 1952 without forming any rapport, wrote of him long after his death:

> Remoteness, a relative absence of intimate personal relationships, is ... a genuine ingredient of certain types of genius. It is certainly true of Einstein, who was himself aware of his absence of contact with human beings; although in his case this certainly did not take the form of a desire for power or glory.

Hence Einstein's departure from Oxford in 1933 for good, despite the best efforts of Lindemann to cajole him back from America in 1934–36. Had he returned, a prestigious professorship in physics, based at Christ Church, would surely have been his for the asking. But it would have entailed the lecturing, committee, political, social and media obligations – not to mention dinner-jackets – that Einstein always regarded as distractions from his science. Even more crucial, Oxford offered no outstanding theoretical physicists with whom Einstein could discuss his ideas (as Schrödinger would discover). In 1933, he wanted to be in a place where he would be absolutely free to think about theoretical physics full-time, either on his own or in conjunction with other leading experts from all over the world – entirely unfettered by non-scientific human demands. Abraham Flexner's newly founded Institute for Advanced Study in Princeton offered him this unique prospect, and with a sufficient salary – unlike an established university such as Oxford or Cambridge.

In *De Profundis*, a gaoled Oscar Wilde remarked: 'Nothing really at any period of my life was ever of the smallest importance to me compared with Art'. From all the available evidence, Einstein felt the same way about Physics.

SOURCES OF QUOTATIONS

NB: The Albert Einstein Archives are at the Hebrew University of Jerusalem: https://albert-einstein.huji.ac.il

Translations from Einstein's German writings held in the Archives are by the author, with some assistance from the staff at the Archives. All other translations are from the published works cited below.

Foreword

6 **England has always produced** Esther Salaman, 'Memories of Einstein', *Encounter*, April 1979, p. 23.

7 **Not even a carthorse** 16 May 1931, in Einstein, 'Diary Notes, April–June 1931, Berlin and Oxford', Albert Einstein Archives, 29–142.

7 **barbarian** Einstein's German poem appeared in translation in *The Times*, 17 May 1955.

Preface

EINSTEIN'S OXFORD BLACKBOARD

8 **I rejoice** The speeches by Shaw and Einstein at the 1930 dinner can be heard on *Albert Einstein: Historic Recordings 1930–1947*, British Library, London, 2005.

9 **two blackboards, plentifully sprinkled** *The Times*, 18 May 1931.

9 **relic of a secular saint** History of Science Museum, Oxford, 'Blackboard Used by Albert Einstein', https://www.hsm.ox.ac.uk/blackboard-used-by-albert-einstein

9 **Some of the scientists** 13 May 1931, in Rhodes Trust Archives, Rhodes House, Oxford, 2694A.

12 **your present** 19 May 1931, in Rhodes Trust Archives, Rhodes House, Oxford, 2694A.

12 **The lecture was indeed well-attended** 16 May 1931, in Einstein, 'Diary Notes, April–June 1931, Berlin and Oxford', Albert Einstein Archives, 29–142.

14 **the holy brotherhood in tails** 2/3 May 1931, in Einstein, 'Diary Notes, April–June 1931, Berlin and Oxford', Albert Einstein Archives, 29–142.

14 **hermit** Einstein's German poem appeared in translation in *The Times*, 17 May 1955.

Chapter 1

OXFORD'S ADVENTURES IN EINSTEIN-LAND

17 **I enclose a note** 14 June 1921, in Cherwell Papers, Nuffield College, Oxford, D53.

17 **They both enjoyed greatly their visit** 15 June 1921, in Cherwell Papers, Nuffield College, Oxford, D53.

17 **The wonderful experiences in England** 21 June 1921, in Albert Einstein, *Collected Papers of Albert Einstein*, vol. 12, Princeton University Press, Princeton, NJ, 2009, doc. 155, p. 113.

19 **He was a man of intuition** Entry on Lindemann, *Oxford Dictionary of National Biography*, 3 January 2008.

19 **The Prof., so it went** R.F. Harrod, *The Prof: A Personal Memoir of Lord Cherwell*, Macmillan, London, 1959, p. 48.

19 **He was an out-and-out inequalitarian** Entry on Lindemann, *Oxford Dictionary of National Biography*, 3 January 2008.

20 **Like many scientists Einstein** *Daily Telegraph*, 22 April 1955.

20 **It has often been asked** Adrian Fort, *Prof: The Life of Frederick Lindemann*, Jonathan Cape, London, 2003, p. 200.

20 **It is a disaster** *Report of Royal Commission on Oxford and Cambridge Universities*, quoted in Robert Fox, 'Einstein in Oxford', *Notes and Records* [of the Royal Society], 9 May 2018, p. 2.

21 **The university and the trustees** 29 June 1927, in Albert Einstein Archives, 32–654.

21 **How gladly would I accept** Undated but probably July 1927, in Cherwell Papers, Nuffield College, Oxford, D54.

22 **During the holidays** 28 August 1927, in Cherwell Papers, Nuffield College, Oxford, D54.

23 **an account of his attempt to derive** *Nature*, 16 May 1931, p. 765.

25 **the discourse should be in English** Einstein's remark was recalled by Mrs K. Haldane in the *Manchester Guardian*, 28 April 1955, soon after Einstein's death.

26 **who have helped me** 'On the Method of Theoretical Physics', in Paul Kent, *Einstein in Oxford: Celebrating the Centenary of the 1905 Publications*, Oxford Physics, Oxford, 2005, p. 12.

27 *Atque utinam Mercurius hodie adesset* *Oxford University Gazette*, 3 June 1931, p. 627.

27 **The doctrine which he interprets** *Oxford Times*, 29 May 1931.

29 **serious, but not wholly accurate** 23 May 1931, in Einstein, 'Diary Notes, April–June 1931, Berlin and Oxford', Albert Einstein Archives, 29–142.

29 **I had noticed his face lit up** Margaret Deneke, 'Professor Albert Einstein', in Deneke Papers, Lady Margaret Hall, Oxford, [MS] p. 4.

29 **and many of our best friends failed** Ibid., [MS] p. 26.

Chapter 2

PHYSICS, MATHEMATICS AND THE NATURE OF REALITY

30 **threw himself into all the activities** Lindemann to Lothian, 27 June 1931, in Cherwell Papers, Nuffield College, Oxford, D56.

31 **very clever man** 5 May 1931, in Einstein, 'Diary Notes, April–June 1931, Berlin and Oxford', Albert Einstein Archives, 29–142.

31 **showed that Milne's distinctive approach** Christopher Smeenk, 'Einstein's Role in the Creation of Relativistic Cosmology', in Michel Janssen and Christoph Lehner (eds), *The Cambridge Companion to Einstein*, Cambridge University Press, Cambridge, 2014, p. 253.

32 *l'affaire* **Einstein** Chapman to Lothian, 10 June 1931, in Rhodes Trust Archives, Rhodes House, Oxford, 2694(2).

32 **he had since discovered** Einstein to Carleton Allen, 9 June 1933, in Rhodes Trust Archives, Rhodes House, Oxford, 2694A.

32 **Theory fed on observation** Corey S. Powell, *God in the Equation: How Einstein Transformed Religion*, Free Press, New York, NY, 2003, p. 97.

34 **a spacious castle in the air** Quoted in Christopher Smeenk, 'Einstein's Role in the Creation of Relativistic Cosmology', in Michel Janssen and Christoph Lehner (eds), *The Cambridge Companion to Einstein*, Cambridge University Press, Cambridge, 2014, p. 241.

34 **not justified by our actual knowledge** Quoted in Ronald W. Clark, *Einstein: The Life and Times*, pbk edn, HarperCollins, New York, NY, 2011, p. 269.

35 **primeval atom** The concept was defined by Georges Lemaître in 'The Beginning of the World from the Point of View of Quantum Theory', *Nature*, 9 May 1931, p. 706.

35 **totally abominable** Quoted in Christopher Smeenk, 'Einstein's Role in the Creation of Relativistic Cosmology', in Michel Janssen and Christoph Lehner (eds), *The Cambridge Companion to Einstein*, Cambridge University Press, Cambridge, 2014, p. 255.

35 **Many theorists saw the phenomenon** Cormac O'Raifeartaigh, 'Einstein's Steady-State Cosmology', *Physics World*, September 2014, pp. 30–33.

36 **It appears that Einstein stumbled** Cormac O'Raifeartaigh, 'Einstein's Blackboard and the Friedmann-Einstein Model of the Cosmos', https://coraifeartaigh.wordpress.com/2015/12/22/einsteins-blackboard/

36 **untrue** Einstein to Carleton Allen, 9 June 1933, in Rhodes Trust Archives, Rhodes House, Oxford, 2694A.

36 **The sanctuary from personal turmoil** Introduction to Michel Janssen and Christoph Lehner (eds), *The Cambridge Companion to Einstein*, Cambridge University Press, Cambridge, 2014, p. 20.

36 **a reckless overestimation** Albrecht Fölsing, *Albert Einstein: A Biography*, Penguin, London, 1998, p. 561.

36 **perhaps the clearest and most revealing expression** Abraham Pais, *Einstein Lived Here*, Oxford University Press, Oxford, 1994, p. 55.

36 **If you wish to learn** 'On the Method of Theoretical Physics', in Paul Kent, *Einstein in Oxford: Celebrating the Centenary of the 1905 Publications*, Oxford Physics, Oxford, 2005, p. 12.

37 **But this is the common fate** Ibid., p. 13.

37 **We honour ancient Greece** Ibid., p. 13.

38 **Pure logical thinking** Ibid., pp. 13–14.

38 **the first creator ... as the ancients dreamed.** Ibid., pp. 15–17.

40 **A far as the laws of mathematics** Einstein, *The Ultimate Quotable Einstein*, ed. Alice Calaprice, Princeton University Press, Princeton, NJ, 2011, p. 371.

41 **a team of scholars ... a unified theory of gravity and electromagnetism.** Graham Farmelo, *The Universe Speaks in Numbers: How Modern Maths Reveals Nature's Greatest Secrets*, Basic Books, New York, NY, 2019, pp. 77–8. The comment by Jeroen von Dongen was made in a personal communication to Farmelo in 2017.

41 **after having ruefully returned** Einstein, *Ideas and Opinions*, ed. Carl Seelig, Three Rivers Press, New York, NY, 1982, pp. 289–90.

Chapter 3

ARTISTIC INTERLUDES

43 **The greatest scientists are artists** Einstein, *The Ultimate Quotable Einstein*, ed. Alice Calaprice, Princeton University Press, Princeton, NJ, 2011, p. 379.

43 **Mozart's music is so pure** Ibid., p. 239.

43 **a fat giant with a red face** 21 May 1931, in Einstein, 'Diary Notes, April–June 1931, Berlin and Oxford', Albert Einstein Archives, 29–142.

44 **more Egyptian than Greek** 26 May 1931, in Einstein, 'Diary Notes, April–June 1931, Berlin and Oxford', Albert Einstein Archives, 29–142.

46 **Generations of Oxford undergraduates** Entry on Helena Deneke, *Oxford Dictionary of National Biography*, 23 September 2004.

46 **In he came with short quick steps** Margaret Deneke, 'Professor Albert Einstein', in Deneke Papers, Lady Margaret Hall, Oxford, [MS] pp. 1–3.

47 **Lady Wylie thought** Ibid., [MS] pp. 4–5.

48 **The guests hastily left the room** 15 May 1931 in Einstein, 'Diary Notes, April–June 1931, Berlin and Oxford', Albert Einstein Archives, 29–142.

48 **Professor Einstein thought we had discovered** Margaret Deneke, 'Professor Albert Einstein', in Deneke Papers, Lady Margaret Hall, Oxford, [MS] p. 4.

48 **The gate at Christ Church is locked** Ibid., p. 14.

49 **He scrutinized the violin** Ibid., pp. 13–14.

50 **relatively good** Quoted in Peregrine White, 'Albert Einstein: The Violinist', *Physics Teacher*, 43, May 2005, p. 287.

50 **the denizen of dimly lit music rooms** Ibid., p. 288.

50 **Professor Einstein was still at his breakfast table** Margaret Deneke, 'Professor Albert Einstein', in Deneke Papers, Lady Margaret Hall, Oxford, [MS] pp. 23–4.

Chapter 4

A STUDENT OF CHRIST CHURCH AND OXFORD

54 **I was shocked** Margaret Deneke, 'Professor Albert Einstein', in Deneke Papers, Lady Margaret Hall, Oxford, [MS] p. 6.

55 **Dundas lets his rooms decay** *The Times*, 17 May 1955.

58 **He was by nature a rebel** Banesh Hoffmann with Helen Dukas, *Albert Einstein: Creator and Rebel*, Viking, New York, NY, 1972, p. 248.

58 **Evening club meal in dinner-jacket** 1 May 1931, in Einstein, 'Diary Notes, April–June 1931, Berlin and Oxford', Albert Einstein Archives, 29–142.

58 **Silent existence in the hermitage** 2/3 May 1931, in Einstein, 'Diary Notes, April–June 1931, Berlin and Oxford', Albert Einstein Archives, 29–142.

60 **was a charming person** R.F. Harrod, *The Prof: A Personal Memoir of Lord Cherwell*, Macmillan, London, 1959, p. 47.

60 **Dr Einstein, do tell me** Arnold Toynbee, *Acquaintances*, Oxford University Press, Oxford, 1967, p. 268. According to Toynbee, Einstein answered Gilbert Murray in Tom Quad: 'I am thinking that, after all, this is a very small star'. However, in a remarkably similar anecdote referring to Murray's house recounted by Murray himself, Einstein answered as I have given: see Otto Nathan and Heinz Norden (eds), *Einstein on Peace*, Schocken, New York, NY, 1960, p. 70.

61 **Your kind letter has filled me** 6 July 1931, in Cherwell Papers, Nuffield College, Oxford, D56.

61 **harmonious community life** 29 October 1931, in Einstein Papers, Christ Church, Oxford.

61 **I cannot help thinking** 24 October 1931, in Einstein Papers, Christ Church, Oxford.

62 **I think that in electing Einstein** 24 October 1931, in Einstein Papers, Christ Church, Oxford.

62 **a German who has no connection** 2 November 1931, in Einstein Papers, Christ Church, Oxford.

62 **The world at large** Quoted in *Financial Times*, 2 February 2008.

63 **excellent institution** 22 May 1931, in Einstein, 'Diary Notes, April–June 1931, Berlin and Oxford', Albert Einstein Archives, 29–142.

63 **pitiful** 23 May 1931, in Einstein, 'Diary Notes, April–June 1931, Berlin and Oxford', Albert Einstein Archives, 29–142.

64 **There are so many fictitious peace societies** Quoted in A. Fenner Brockway, *New World*, July 1931, in Otto Nathan and Heinz Norden (eds), *Einstein on Peace*, Schocken, New York, NY, 1960, p. 140.

64 **Even if only two per cent** Otto Nathan and Heinz Norden (eds), *Einstein on Peace*, Schocken, New York, NY, 1960, p. 117.

64 **For example, he suggested** *The Friend*, 12 June 1931, p. 567.

67 **The types of humanity** Margaret Deneke, 'Professor Albert Einstein', in Deneke Papers, Lady Margaret Hall, Oxford, [MS] p. 7.

67 **Make sure you get home** 9 May 1931, quoted in Einstein, 'Diary Notes, April–June 1931, Berlin and Oxford', Albert Einstein Archives, 29–142.

67 **Soon there will be water-closets** 25 May 1931, in Einstein, 'Diary Notes, April–June 1931, Berlin and Oxford', Albert Einstein Archives, 29–142.

67 **Long-branched and delicately strung** Quoted in Walter Isaacson, *Einstein: His Life and Universe*, Simon & Schuster, London, 2007, p. 361.

68 **tiny moustached and hatted figure** William Golding, 'Glimpses', *Reader's Digest*, August 1968, p. 182, and extended version in Golding, 'Thinking as a Hobby', http://www.brunswick.k12.me.us/hdwyer/thinking-as-a-hobby-by-william-golding/

Chapter 5

REFUGEE FROM NAZISM

69 **The movement to induce Prof. Einstein** *Jewish Telegraphic Agency*, 26 December 1930.

69　**when we clean house** Quoted in Ronald W. Clark, *Einstein: The Life and Times*, pbk edn, HarperCollins, New York, NY, 2011, p. 543.

70　**grand** 1 May 1933, in Cherwell Papers, Nuffield College, Oxford, D57.

70　**there was not much prospect** 4 May 1933, in Cherwell Papers, Nuffield College, Oxford, D57.

72　**one room** 7 May 1933, in Cherwell Papers, Nuffield College, Oxford, D57.

73　**You are not a father yourself** 9 May 1933, in Cherwell Papers, Nuffield College, Oxford, D57.

75　**You know, I think, that I have never** Max Born and Albert Einstein, *The Born–Einstein Letters*, 2nd edn, Macmillan, London, 2005, pp. 111–12.

75　**a poor forlorn little figure** Quoted in Ronald W. Clark, *Einstein: The Life and Times*, pbk edn, HarperCollins, New York, NY, 2011, p. 582.

76　**I wish to preface what I have to say** 'On the Method of Theoretical Physics', in Paul Kent, *Einstein in Oxford: Celebrating the Centenary of the 1905 Publications*, Oxford Physics, Oxford, 2005, p. 12.

77　**intellectual and individual freedom** 'Professor Einstein' in [Albert Hall], *Royal Albert Hall: Tuesday October 3rd 1933*, London, 1933 (souvenir booklet containing speeches by Einstein and others), p. 5. No publisher is given; the booklet was presumably produced by Oliver Locker-Lampson. See film of Einstein talking at: https://www.youtube.com/watch?v=ZBage5Ff57E

77　**I had never seen him before** Notes by William Beveridge for his BBC talk, 'Professor Albert Einstein at the Albert Hall', 3 October 1933, in Beveridge Papers, London School of Economics, 9A/45/4.

77　**I could not believe that it was possible** *Eastern Daily Press*, 4 October 1933. Einstein's comment appears to have been made to the reporter, who is unnamed.

Postscript

EVERYWHERE AND NOWHERE AT HOME

80　**hermit** *The Times*, 17 May 1955.

81　**a wanderer on the face of the earth** *Eastern Daily Press*, 4 October 1933.

81　**Remoteness** Letter to Anthony Storr, 29 September 1978, in Isaiah Berlin, *Affirming: Letters 1975–1997*, eds Henry Hardy and Mark Pottle, Chatto & Windus, London, 2015, p. 84.

82　**Nothing really at any period of my life** Quoted in Andrew Robinson, *Genius: A Very Short Introduction*, Oxford University Press, Oxford, 2011, p. 77.

FURTHER READING

[Albert Hall], *Royal Albert Hall: Tuesday October 3rd 1933*, [publisher unknown], London, 1933 (souvenir booklet containing speeches by Einstein and others)

Born, Max and Albert Einstein, *The Born-Einstein Letters*, 2nd edn, Macmillan, London, 2005

Calaprice, Alice, Daniel Kennefick and Robert Schulmann, *An Einstein Encylopedia*, Princeton University Press, Princeton, NJ, 2015

Clark, Ronald W., *Einstein: The Life and Times*, pbk edn, HarperCollins, New York, NY, 2011

Clary, David C., *Schrödinger in Oxford*, World Scientific, London, 2022

Einstein, Albert:

  Ideas and Opinions, ed. Carl Seelig, Three Rivers Press, New York, NY, 1982

  The Collected Papers of Albert Einstein, vols 1–17 (various editors), Princeton University Press, Princeton, NJ, 1987–

  Relativity: The Special and the General Theory, trans. Robert W. Lawson, Routledge Classics, London, 2001

  The Ultimate Quotable Einstein, ed. Alice Calaprice, Princeton University Press, Princeton, NJ, 2011

Eisinger, Josef, *Einstein on the Road*, Prometheus, New York, NY, 2011

Farmelo, Graham, *The Universe Speaks in Numbers: How Modern Maths Reveals Nature's Greatest Secrets*, Faber & Faber, London, 2019

Fölsing, Albrecht, *Albert Einstein: A Biography*, Penguin, London, 1998

Fort, Adrian, *Prof: The Life of Frederick Lindemann*, Jonathan Cape, London, 2003

Fox, Robert, 'Einstein in Oxford', *Notes and Records*, 9 May 2018, pp. 1–26

Fox, Robert and Graeme Gooday (eds), *Physics at Oxford 1839–1939: Laboratories, Learning, and College Life*, Oxford University Press, Oxford, 2005

Golding, William, 'Glimpses', *Reader's Digest*, August 1968, p. 182

Harrod, R.F., *The Prof: A Personal Memoir of Lord Cherwell*, Macmillan, London, 1959

Hoffmann, Banesh with Helen Dukas, *Albert Einstein: Creator and Rebel*, Viking,

New York, NY, 1972

Isaacson, Walter, *Einstein: His Life and Universe*, Simon and Schuster, London, 2007

Janssen, Michel and Christoph Lehner (eds), *The Cambridge Companion to Einstein*, Cambridge University Press, Cambridge, 2014

Janssen, Michael and Jürgen Renn, 'Einstein Was No Lone Genius', *Nature*, 19 November 2015, pp. 298–300

Kent, Paul, *Einstein in Oxford: Celebrating the Centenary of the 1905 Publications*, Oxford Physics, Oxford, 2005

Lindemann, Frederick A. (Lord Cherwell), 'The Genius of Einstein', *Daily Telegraph*, 22 April 1955

Morrell, Jack, *Science at Oxford 1914–1939: Transforming an Arts University*, Oxford University Press, Oxford, 1997

Murray, Gilbert, *An Unfinished Autobiography*, Jean Smith and Arnold Toynbee (eds), George Allen and Unwin, London, 1960

Nathan, Otto and Heinz Norden (eds), *Einstein on Peace*, Schocken, New York, NY, 1960

O'Raifeartaigh, Cormac, 'Einstein's Steady-state Cosmology', *Physics World*, September 2014, pp. 30–33

Pais, Abraham, *Einstein Lived Here*, Oxford University Press, Oxford, 1994

Powell, Corey S., *God in the Equation: How Einstein Transformed Religion*, Free Press, New York, NY, 2003

Robinson, Andrew:

> *Einstein: A Hundred Years of Relativity*, Princeton University Press, Princeton, NJ, 2015

> 'Einstein in Oxford', *Christ Church Matters*, 2016, pp. 36–7

> 'Einstein Said That – Didn't He?', *Nature*, 3 May 2018, p. 30

> 'Einstein in Oxford', *Physics World*, June 2019, pp. 32–6

> *Einstein on the Run: How Britain Saved the World's Greatest Scientist*, Yale University Press, London, 2019

Rosenkranz, Ze'ev, *The Einstein Scrapbook*, John Hopkins University Press, Baltimore, MD, 2002

Rowe, David E. and Robert Schulmann (eds), *Einstein on Politics: His Private Thoughts and Public Stands on Nationalism, Zionism, War, Peace, and the Bomb*, Princeton University Press, Princeton, NJ, 2007

Toynbee, Arnold, *Acquaintances*, Oxford: Oxford University Press, 1967

White, Peregrine, 'Albert Einstein: The Violinist', *Physics Teacher*, May 2005, pp. 286–8

ACKNOWLEDGEMENTS

I dedicate this book to my old school friend James Lawrie, who in 2015 organized a talk for me at his college, Christ Church, on Einstein in Oxford – which first gave me the idea of researching this subject in depth. Thanks also to the Albert Einstein Archives in Jerusalem and to the fellows of Christ Church, Lady Margaret Hall and Nuffield College in Oxford, as well as the Rhodes trustees, for permission to quote correspondence relating to Einstein's visits to Oxford, notably the diary of Margaret Deneke and the letters of Frederick Lindemann (Lord Cherwell). Matin Durrani encouraged me by publishing my article, 'Einstein in Oxford', in *Physics World*, as did Silke Ackermann by inviting me to speak on the subject at the History of Science Museum in Oxford. I am also grateful to Samuel Fanous, head of publishing at the Bodleian Library, for his sustained interest, and to the editorial staff: Susie Foster, Miranda Harrison, Dot Little, Janet Phillips and Leanda Shrimpton. I hope the book explains the enduring fascination of Einstein's poem about himself, written in Christ Church in 1931 and now kept in the Bodleian Library.

PICTURE CREDITS

INDEX